The Struggle for Food Sovereignty

The Struggle for Food Sovereignty

Alternative Development and the Renewal of Peasant Societies Today

Edited by Rémy Herrera and Kin Chi Lau

With Samir Amin, Gérard Choplin et al., Sam Moyo, Utsa Patnaik, Jade Tsui Sit, João Pedro Stedile, Poeura Tetoe and Erebus Wong

PLUTO PRESS

First published 2015 by Pluto Press
345 Archway Road, London N6 5AA

www.plutobooks.com

British Library Cataloguing in Publication Data
A catalogue record for this book is available from the British Library

ISBN   978 0 7453 3595 7   Hardback
ISBN   978 0 7453 3594 0   Paperback
ISBN   978 1 7837 1505 3   PDF eBook
ISBN   978 1 7837 1507 7   Kindle eBook
ISBN   978 1 7837 1506 0   EPUB eBook

10   9   8   7   6   5   4   3   2   1

Typeset by Stanford DTP Services, Northampton, England
Text design by Melanie Patrick

CONTENTS

WORLD FORUM
FOR ALTERNATIVES

The World Forum for Alternatives (WFA) is a network of individuals and institutions committed to a progressive and anti-imperialist perspective, supporting the aspirations of the nations of the 'South'– i.e. the dominated peripheries in the global capitalist system – to become equal to the 'North' in all the dimensions of life: economic development, social welfare, political independence and cultural respect. WFA members are intellectuals, sharing modest but nonetheless ambitious targets: modest because they do not consider themselves 'leaders' of progressive social and political forces in struggle, and ambitious in the sense that they do feel that they are among those who can provide in-depth analyses of the realities and challenges, and can thus be 'useful' for the progressive forces in struggle. They also allow themselves to suggest strategies for action and are keen to see those analyses and suggestions move out of 'restricted' teams of thinkers to reach the political and social progressive forces.

Since its creation in Cairo in 1997, after a first meeting in Louvain-la-Neuve in 1996, the WFA has conducted activities in accordance with its purpose and platform. Some of its main activities have included the 'Anti-Davos in Davos' (in January 1999, a media event which gave visibility to the network); the Bamako General Assembly (in January 2006, attended by around 200 members of the Enlarged Council of the WFA); and the Caracas General Assembly (in October 2008, attended by 250 members of the Enlarged Council). A great number of other events have been organised or co-organised by the WFA, such as round tables in the successive rounds of the World Social Forum (Porto Alegre, Mumbai, Nairobi, Dakar and Tunis), as well as in other continental, regional and national social forums. Numerous activities have also been co-organised by the WFA in association with various partners in Africa, Asia, Latin America and Europe.

What we might call the 'WFA political platform' has been formulated, in particular, in the 2006 Bamako Appeal (available in several languages). This has inspired many progressive think tanks. In this spirit, the WFA has organised much debate during the last 15 years in several key areas on such challenges as building the unity of the labouring classes; constructing peasant perspectives for the half of humankind still living in rural areas; associating the democ-

, and finding new alternatives in the ongoing phase of the crisis of the global capitalist system.

Among the leading members of the WFA are: S. Amin (Egypt, France), A.K. Bagchi (India), P. Beaudet (Canada), A. Buzgalin (Russia), B. Cassen (France), H. Chalbi-Drissi (Morocco), A. Conchiglia (Italy), Dai J. (China), A. Dansoko (Senegal), W. Dierckxsens (The Netherlands), B. Founou-Tchuigoua (Cameroon), J.B. Foster (USA), P. Gonzalez Casanova (Mexico), M. Habashi (Egypt), R. Herrera (France), F. Houtart (Belgium), Huang P. (China), V.H. Jijon (Ecuador), B. Kagarlisky (Russia), A. El Kenz (Algeria), M. Katsumata (Japan), Z. Kowalewski (Poland), Lau K.C. (China), I. Lindberg (Sweden), H. Marais (South Africa), I. Monal (Cuba), S. Moyo (Zimbabwe), P.K. Murthy (India), K. Mushakoji (Japan), P. Nakatani (Brazil), I. Rauber (Argentina), F. Rochat (Switzerland), I. Shivji (Tanzania), H. Shaarawi (Egypt), C. Tablada (Cuba), A. Tujan (Filipina), Wang H. (China), Wen T. (China).

ABBREVIATIONS

AGRA   Alliance for Green Revolution in Africa
AusAID   Australian Agency for International Development
CAP   Common Agricultural Policy
CLOC   Latin American Coordinating Committee of Rural Organisations
COPA   Committee of Professional Agricultural Organisations
CPC   Communist Party of China
CPE   European Farmers' Coordination
DRC   Democratic Republic of Congo
ECVC   European Coordination Via Campesina
EPA   Economic Partnership Agreements
FAO   Food and Agriculture Organization of the United Nations
FDI   Foreign direct investment
FTLRP   Fast Track Land Reform Programme
GM[O]   Genetically modified [organism]
IALA   Latin American Institute of Agroecology
IFAP   International Federation of Agricultural Producers
IIRR   International Institute of Rural Reconstruction
IMF   International Monetary Fund
ISI   Import substitution industrialisation
LSCF   Large-scale commercial farming
MIJARC   International Movement of Catholic Agricultural and Rural Youth
SACU   Southern African Customs Union
SADC   Southern African Development Community
SAPs   Structural adjustment plans
SCTL   Civil Society for Land in Larzac
SEPA   State Environmental Protection Agency
SEZ   Special economic zone
USAID   US Agency for International Development
WFA   World Forum for Alternatives
WTO   World Trade Organization

INTRODUCTION

Family Agriculture in the Present World:
Regional Perspectives

Rémy Herrera and Kin Chi Lau

This book, driven by a collective reflection within the framework of the World Forum for Alternatives, is dedicated to the problems faced by Southern and Northern family farms in the current neoliberal era of financial capital domination worldwide, and to the revival of peasant struggles for their social emancipation and legitimate right of access to land and food. Obviously, such struggles also concern all categories of workers and the people as a whole because what is at stake is the challenge to reach food sovereignty and to build our societies, at the local, national and global levels, on the principles of social justice, equality and real democracy.

The food and agricultural crises, which erupted in 2007–08 and resulted in catastrophic effects on the peoples of numerous countries of the South, especially Africa, as well as popular rebellions, represent two of the many dimensions of the crisis of the capitalist world system. Other very worrying aspects include socio-economic, political and ideological ones, energy and climatic ones. The food and agricultural dimensions of the current systemic crisis reveal the global failure and deep dysfunctions that characterise the agricultural 'model' imposed worldwide by financial capital and transnational agribusiness corporations since the beginning of the neoliberal era in the late 1970s, along with the implementation of austerity policies in the North and the structural adjustment plans (SAPs) in the global South. For more than three and a half decades the peasantries of the world have been suffering an intensification of attacks by capital on their land, natural resources and means of production. These attacks have also been eroding national sovereignty and the role of the state, destroying individuals, families and communities, devastating the environment, and threatening the survival of huge numbers of human beings across the world.

The dysfunctions affecting the agricultural sectors can be perceived by identifying a series of striking paradoxes. As a matter of fact, approximately three billion people on the planet today continue to suffer from hunger (one-third) or malnutrition (two-thirds), although agricultural production greatly exceeds food needs, with an effective overproduction of at least 150 per cent. Furthermore, a huge majority of these people are themselves peasants or living in rural areas: three-quarters of those suffering from undernourishment are rural. Meanwhile, the expansion of the areas for cultivation worldwide is accompanied by a significant decline in peasant populations compared to the populations in the urban areas, which absorb the massive and persistent rural exodus, mainly into growing miserable slums. Moreover, an increasing proportion of land is cultivated by transnational corporations which do not direct their agricultural production towards food consumption, but rather towards energy or industrial outlets (for example, agro-fuels). In most countries of the South that are excluded from the benefits of capitalist globalisation, particularly sub-Saharan Africa and South-East Asia, a relative dynamism of agricultural exports derived from rental commercial crops coexists with increasing imports of basic products to meet food needs. Clearly, and urgently, things must change.

This project was initiated as an attempt to make sense of how these urgent global problems are manifested in the North and the South, and while there are common traits in how global capital goes after profit, the effects on the ground differ. Hence, it is important for struggles in different parts of the world – affected differently but also sharing related features – to develop a concerted understanding of the problems and prioritise strategies that take heed of the differences, and to share common visions for the future. Thus, in this book, authors from different continents have been invited to make their contributions and offer different perspectives and reflections and to relate their local struggles and immediate concerns to a global and long-term vision.

Theoretical and Historical Framework

The first chapter provides a broad theoretical and historical framework for the book. Samir Amin proposes a series of analytical elements to answer major questions about the appropriate kind of agriculture (capitalist, socialist or peasant) to guarantee the objective of food sovereignty; the agricultural productions to be prioritised to reach a development model, which is able to conciliate the improvement in food supply and the preservation of the environment for the generations to come; and to reflect on the resolution of the agrarian question by constructing convergence of struggles within diversity.

First, he analyses family agriculture in the present world and the differences between the North and the South. In the North (North America and Western Europe), a modern and highly productive family agriculture largely dominates, absorbs technological innovations, efficiently supplies these countries' food demands and produces exportable surpluses. However, while totally integrated into the capitalist system, this agriculture does not share a key characteristic of capitalism: its labour organisation generally requires a reduced and polyvalent workforce. Furthermore, within the capitalist logic, a significant part of the income generated by farmers – even when they own land and equipment and receive subsidies – is controlled and collected by segments of commercial, industrial and financial capital, implying that their remuneration does not correspond to their productivity. Therefore, family agriculture can be assimilated to the status of a subcontractor or an artisan working in a putting-out system, and squeezed between supermarkets, agribusiness and banking.

In the South where peasant families constitute almost half of humanity (three billion people), the types of agricultural systems vary widely, with considerable differences in productivity among them (from mechanised latifundia to small or micro-parcels, with lands for self-consumption or cash crop exports, etc.). But, taken as a whole, these Southern farms – which are more often than not peasant ones – suffer from a huge and growing productivity gap as compared to those of the North. Most family farms of the South are under-equipped, non-competitive and destined for subsistence food, which explains the poverty of the rural world, the inefficiency to supply food to cities, and other serious problems affecting these societies. However, the Southern peasant agriculture is also largely integrated into the local and dominant global capitalist system and their profits are consequently siphoned off by dominant capital.

Here, the crucial question is whether agriculture in the South could be modernised by capitalism. Amin says no, and demonstrates why this is so. He criticises the notion of 'food security' – as an alternative to food sovereignty – disseminated by international organisations and Northern governments, according to which the South should rely on a specialisation in cash-crop products for export to cover food deficits. This results in disaster, as the recent food crisis has shown. What is absolutely necessary is food sovereignty. For that, a sine qua non is access to land for all peasants, to be considered a goal towards which most struggles in rural areas are oriented. For this reason Amin differentiates the types of land tenure systems in the South, depending on the ownership status.

The first system is land tenure based on private ownership – 'absolute right', only limited by public laws and eventual environmental regulations. Since the 'enclosures' process in early capitalism in Western Europe, this is seen as the 'modern' form of landownership by the 'liberal' ideology's rhetoric and management rationale by making land a 'merchandise' exchangeable at market price. Opposing this idea, Samir Amin asserts that it is unsustainable to draw from the construction of Northern modernity rules for the advancement of the peoples of the global South. To change land into private property, the present reactivation of the 'enclosures' process involves dispossession of peasants, as in the colonial times. Other forms of regulating the right to use land are conceivable and can produce similar results, avoiding the foreseeable destruction by capitalism.

Land tenure not based on private ownership is the second system, which takes heterogeneous forms and where access to land is simultaneously regulated by various rules that are derived from institutions involving individuals, communities and the state. Among these are 'customary' rules that traditionally guarantee access to land to all families – but this does not mean equal rights. These rights of use by communities are limited by the state and only exist today in deteriorated forms, attacked by capitalist expansion and its associated private appropriation. Amin gives several past and present examples of such situations in Asia and Africa. Frequently, European colonial powers left customary practices alone, allowing them to retain their domination (like 'économie de traite' in the French colonial administration). The same phenomenon is occurring today under imperialist pressures.

However, popular revolutions in Asia or Africa sometimes challenged this legacy. Among them, China and Vietnam (we might add Cuba in Latin America to this list too) constitute unique examples of the success of a land system based on the rights of all peasants within the village. This constitutes equal access to and use of land, with the state as the sole owner and equal land distribution among usufructuary peasant families. Amin examines the evolution of this system based on the suppression of private landownership, up until the present times, as well as its viability and ability to resist the attacks it is suffering in rural China and Vietnam nowadays. Peasant struggles are currently active in these two countries to defend the most precious accomplishment of their revolutions.

Elsewhere, agrarian reforms implemented by non-revolutionary hegemonic blocks generally only dispossessed large landowners to the benefit of middle (or even rich) peasants, ignoring the interests of the poor. However, Samir Amin affirms that new waves of agrarian reforms are needed today to meet the legitimate demands of the poorest and landless peasants in India, South-East Asia, Kenya, South Africa, the Arab countries and many parts of Latin America. This is true even for other Southern regions where capitalist private ownership rights have not yet penetrated deeply (or formally), such as in inter-tropical Africa.

This could be done through an expansion of the definition of public property to include land, along with a movement of democratisation (and not 'retreat') of the state and the minimisation of inequalities. Nevertheless, the success of these agrarian reforms always remains uncertain, because such redistributions maintain tenure systems led by the principle of ownership and even reinforce the adherence to private property. In the dominant discourse, serving the interests of capital and its agribusiness model, a 'modern reform' of the land tenure system means privatisation, which is the exact opposite of what is actually required by the challenges of building democratic and alternative agricultural projects based on prosperous peasant family economies as a whole. Consequently, the only obstacle to the fast trend of commodification and private appropriation of landownership is the resistance and organisation of its victims: the peasants.

Regional Perspectives

The following parts of the book present and analyse, by region, the experiences of peasant struggles to defend their inalienable rights for access to land and food sovereignty. The regions covered are Latin America, Africa, Asia, Oceania and Europe.

In Chapter 2, João Pedro Stedile examines the forms and tendencies of capital penetration in the agricultural sector in Latin America, especially through transnational corporations. Stedile also studies the current challenges imposed on peasant movements of this continent and their programmes, in particular those of the international movement La Via Campesina.

Stedile begins by analysing the mechanisms through which capital accumulated outside of agriculture has taken control of this sector and concentrated it worldwide in the current phase of financialised capitalism. Discussing the consequences of the recent capital crisis and the intensified assault of financial capital on agriculture and the environment, Stedile elaborates how, due to the crisis, large Northern corporations fled to peripheral countries to save their volatile capital by investing in fixed assets, such as land, minerals, raw materials, water, biodiversity territories, or tropical agriculture, and by taking over renewable energy sources, particularly productions of sugar cane and maize for ethanol or soybean and African palm for vegetable oil (agro-fuels). This generated huge speculative operations in the futures markets and a rise in the prices of agricultural (and mining) goods traded in the global futures stock exchange markets, without any correlation to production costs and the actual value of the socially needed labour time.

Stedile then analyses the consequences of the imposition of corporate private ownership of natural resources on the life and organisation of the peasants, with peoples and states losing sovereignty over food and productive processes. The destructive 'model' of capital for agriculture – agribusiness, or 'agriculture without people' – brings deep and insuperable contradictions that need to be understood in order to act upon them.

With this aim, Stedile defends what could be the main elements of a peasant programme that promotes workers' control, anti-capitalist agriculture, food sovereignty and environmental protection in the countries of the South where the peasantry predominates and suffers. This alternative platform, promoted by La Via Campesina, among others movements, includes: prioritising policies of food sovereignty and healthy foods; preventing the concentration of private land and nature ownership; diversifying agriculture; increasing labour and land productivity and adopting machines that respect the environment; reorganising agricultural industries into small- and medium-scale units controlled by workers and peasants; controlling food production by domestic social forces and prohibiting foreign capital from owning land in any country; stopping deforestation; preserving and disseminating native improved seeds and preventing the spread of genetically modified seeds (GMOs); ensuring access to water as the right to a common good for every citizen and developing infrastructure in rural communities; implementing a popular energy sovereignty and reviewing current models of transportation; ensuring the rights of indigenous communities; promoting socially oriented public policies for agriculture; universalising social welfare for the entire population; generalising educational (and literacy) programmes in rural areas and enhancing local cultural habits; changing the international free-trade agreements that function to the detriment of the peoples; and encouraging social relations based on human values built over millennia, such as solidarity and equality – which are the very values of socialism.

Stedile presents some organisational and political challenges for peasant movements, at the local and global levels, in order to face the current disad-vantageous balance of power, where global capital is on the offensive to control nature and agricultural goods. Such an analysis results from the experienced realities in Latin America, especially in Brazil, and from the struggles and resistances of these peasant movements against capitalist destruction. And last, Stedile suggests addressing the interests of transnational capital and its control mechanisms by: building a popular, alternative development model of agricultural production managed by the peasants and workers; by transforming the struggle for land into a struggle for territory; developing a technological matrix based on agroecology, free schools in the countryside, training programmes at all levels and alternative means of mass communication; and creating opportunities for mass social struggles and building alliances against the class enemies with all sectors living in rural areas as well as city workers, nationally and internationally.

In Chapter 3, with a specific focus on Southern Africa, Sam Moyo presents an overview of the African peasantries who have suffered repeated attacks under colonialism, post-independence and neoliberal capitalism. He goes on to outline the perspectives of rebuilding them on the reaffirmation of the inalienability of land rights and collective food sovereignty. His starting point is the desperate situation of most African peasants, who are facing a crisis of social reproduction, food insecurity and insufficient incomes from farming, and their survival strategies despite the state's withdrawal. Regardless of the diversity of African agriculture, its persistent and generalised failure to increase productivity and supplies as well as to resolve key agrarian questions of enhancing the social reproduction of the majority of the peasantries – conceived as elements of democratisation and national development – is clear and dramatic.

Centuries of systemic land alienation and exploitation of peasantries' labour, through unequal integration into the capitalist world system during colonial and post-independence periods, has resulted in the underdevelopment of the agrarian systems. SAPs exacerbated extroversion, extraction of surplus value, land concentration, food imports and aid dependency. Recently, a new assault led by foreign land-grabbing actors dispossessed the peasantry of its lands and natural resources and intensified its labour exploitation. Such accumulation processes undermine the social value of peasant production based on self-employed family labour and self-consumption as well as its ability to adopt technologies and crops to expand low energy-intensive production for its social reproduction. These evolutions, which are driven by financialised capital and agribusiness at the expense of the poor and marginalised peasantries, fuel local conflicts and accentuate the polarisation of agrarian accumulation (from 'above' rather than from below).

Moyo examines the long-running history of the destruction of African food production systems by analysing the trajectory of primitive accumulation and disarticulation of these agrarian societies. He describes the various phases, forms and trends of land alienation, dispossession and incorporation of the peasantries, from colonialism, post-independence developmentalism, to neoliberalism and its re-institutionalised primitive accumulation. He finally touches upon the current crisis involving land grabbing and 'contracted farmers'. Then, he explains the underdevelopment of the agrarian productive forces, using examples from country members of the malintegrated Southern African Development Community (SADC), and the persistence of qualitative changes in the agrarian surplus extraction and its externalisation through the unequal world and subregional trade regimes under neoliberalism. Here, the recent global food price and agrarian crisis, especially in the SADC region, as well as South African capital's hegemony are studied. Moyo assesses the social consequences of such processes on the collapse of basic food consumption and the fast increase in food-related poverty – except in a few 'secure' enclaves (in South Africa) – and on the more recent alternative strategies within the neoliberal context and the 'push' to universalise the commodification of land.

Moyo concludes that the real alternative is one that supports priorities given to food sovereignty and a sustainable use of resources by autonomous small producers, in which democracy is inclusive and solidly founded on social progress. This requires a wide range of public policy decisions of restructuring these food systems, including the choices of the basic commodities to be produced in order to satisfy social needs, a redistribution of the means of food production, especially land, inputs and water, substantial infrastructural investments, and enhancing the peasantries' human resources. If the state pursues more systematic and voluntary agrarian reforms to sustain rural development at the national level, this task will also include regional integrations. As a consequence, a reorientation of the SADC region's agricultural (and industrial) policies towards more collective strategies to defend food sovereignty and land rights is needed, in order to reverse the present free-trade and market-based approach of this regionalisation.

Chapter 4 moves to Asia, where Erebus Wong and Jade Tsui Sit, following Wen Tiejun's theses, attempt to rethink the main problematics of 'rural China' in the development of the country in order to argue for rural regeneration as an alternative to a destructive 'modernisation'. The latter is often reduced to industrialisation and the empowering of the state, pursued through several phases from the middle of the nineteenth century to the revolutionary period with its radical social changes. It seems to be relevant to reconsider the intellectual heritage of the rural reconstruction movement – active during the 1920s and 1930s but much neglected today – in post-developmental China, where the rural sector has been historically exploited.

To understand the present situation of China's peasantry – which is the majority of its population – it is necessary to examine in depth the mechanisms involved beyond the collectivisation–liberalisation dichotomy. Land is a key issue for China, which has to nourish 19 per cent of the world's population with 8 per cent of its arable land. In spite of considerable agricultural output, only 13 per cent of its total land area can be cultivated. The explanation is to be found in the fact that land is collectively owned by village communities and distributed within peasant households, who use it mainly for food production to maintain self-sufficiency. Wong and Sit propose a historical overview of China's modernisation to capture the essence of its developmental trajectory in the last 60 years. After 1949, the new regime underwent a period of Soviet-style industrialisation, installing an asymmetric dual system clearly unfavourable to the peasantry. However, despite the industrialisation strategy, the peasantry has benefited from the radical land reforms.

Nowadays, many peasants (and workers) are increasingly suffering from exploitation and injustice, but a few residual socialist practices subsist, including the legacy of land reforms. In the mid 1980s, the promotion of export-oriented growth generated flows of migrant workers from the rural areas to cities – mostly consisting of surplus labour force from rural households that owned a small plot, without land expropriation. The rural sector took up the cost of social reproduction of labour and served as a buffer to absorb social risks in urban areas caused by current pro-capital reforms. It also revealed its stabilising capacity by regulating the labour market and reabsorbing unemployed migrant workers in cities during cyclic crises.

Nevertheless, mainstream intellectuals support the neoliberal ideology to advocate land commodification. Under the pressures of construction projects led by fiscally constrained local governments and real-estate speculators, land expropriation accelerated in the 1990s. Between 40 and 50 million peasants lost their land; the landless appeared in the 2000s, especially after the 2003 law modifying collective arable land legislation and excluding a new generation from land allocation through redistribution. Wong and Sit explain the dangers associated with such evolutions, which weaken the mechanisms of risk management through internalisation in rural community, in a time when 200 million peasant migrant workers are living in cities and evolving into the working class. This is why, inspired by Wen Tiejun's analysis of the agrarian and rural sectors of China, which are considered to have played the role of social stabiliser by absorbing the cost of crisis, they defend collective landownership in rural areas as the most precious legacy of the 1949 revolution.

China's take-off is largely based on the exploitation of its rural sector. Today, the export-oriented model has become such a path-dependency model and internal disequilibriums are so deep that China has to make great efforts to switch its trajectory of development in order to invest into rural society, to guarantee social progress and to preserve the environment. According to the authors, solutions for an alternative path could be to reactivate and revalorise the status of the peasantry, to rediscover the pioneering ideas of the rural reconstruction movements (promoted by Liang Shuming and James Yen, among others), and to support the experiments of rural regeneration currently developed in the country, as renewed and powerful insights, both popular and ecological, to overcome the destructive aspects of contemporary global capitalism.

In Chapter 5, Utsa Patnaik exposes the political–economic context of the peasant struggles for livelihood security and land in India. She begins by recalling that the peasantry and rural workers of the global South are under historically unprecedented pressures today from attacks by capital, especially on the means of securing livelihood – among these is an assault on land – in order to divert its use for capital's own non-agricultural purposes. Such a movement looks similar to that of primitive accumulation in Western Europe between the sixteenth and nineteenth centuries, but today, the Southern peasantry has nowhere to migrate, except to the immense slums of the megalopolis. However, peasants are now turning from passive resistance to active contestation of global capital domination, transforming themselves from objects to subjects of history.

Patnaik examines, in a first part, the agrarian distress, suicides and unemployment in India. She points out that inequalities have increased considerably in the country from the early 1990s under neoliberal policies and that the living condition of the masses of the labouring poor today is globally worse – except where positive interventions have taken place to stabilise livelihood. In rural India, this situation results from attempts to take over peasant lands and resources by domestic and foreign corporations, supported by the state. In parallel, unemployment is partly due to the inability to translate higher economic growth without income redistribution into job creation, while purchasing power has been eroded by the inflated prices of basic needs for ordinary people, forgotten by the ruling classes' strategy of submission to financial capital.

The author points out that the main trend observed in the Indian economy – which has two-thirds of its workforce occupied in agriculture – is that the relative share of agriculture, forestry and fishing in the gross domestic product, especially for key crops like food grains, has declined; industry's share has stagnated, but that of services has increased fast. In a general context of trade openness, fiscal contraction, price-stabilisation system dismantling and land acquisition for special economic zones (SEZs), the state has launched an attack on small farmers, in the name of 'development' but in fact for the benefit of a small minority of real estate speculators, thus creating an agrarian crisis intensifying into the struggle for land.

As a consequence, small producers have been exposed to the ups and downs of prices, have been forced into debt to money lenders and banks, have lost lands against unpaid debts or have even committed suicide. With the implementation of the neoliberal agenda, land ownership concentration is happening at an all-India level and livelihood insecurity is spreading. Therefore, farming is becoming unviable. The author analyses the ongoing resistances of farmers to land acquisition (particularly when the state creates SEZs) or to change in land use (setting up extractives). She describes the repression suffered by peasant rebellions, in Maharashtra and Uttar Pradesh for instance, and also the victories won when the state governments have had to withdraw their projects or concede the right to compensation, as in West Bengal.

Patnaik recalls the fundamental economic characteristics of land, which is not produced by human labour, and the implications of its pricing, which is completely different from that of agricultural commodities (prices are anchored to amounts of labour used for producing them). Based on market capitalisation of incomes, the price of land – in a capitalist system – can vary considerably, depending on its use and the associated yield. Here lies the root of the discontent of farmers, constrained (and cheated) by the state governments to sell their lands at extremely low prices, that is, with compensation far below the profits earned by private investors or speculators (sometimes subsidised), who parcel them for lucrative commercial or residential purposes. One adverse effect (among others, including environmental ones) is that the total cropped area becomes stagnant and the growth in output slows down, leading to inflation in food prices and a contraction of demand. The author finally asserts that to think – like the corporates in collusion with the state do in India – that peasants can be treated as dupes is a mistake; they are now aware of their rights and are strongly resisting their exploitation.

Chapter 6 deals with Oceania, more specifically Papua New Guinea, which gained independence from Australia as recently as 1975. The authors Rémy Herrera and Poeura Tetoe elucidate the 'Papua Niugini paradox', that is, the striking coexistence of an alleged 'archaic' (i.e. not based on private property) system of landownership – as in most Oceanian insular countries – and the vivacity of the peasants' resistance against current neoliberal forms of capitalism, such as the penetration of foreign direct investment in mining, hydrocarbons and natural resources, including forestry and water. Access to land is a real issue in this country where a majority of the population is still involved in subsistence crops for self-consumption, 'customary' rules persist on more than 90 per cent of the soil territory, and the use of land is the source of acute conflicts between transnational corporations, the state and the society.

To begin with, the authors examine the people's attachment to land. European colonisation integrated the indigenous people into global capitalism, transforming most of them into small farmers and making them dependent on colonial plantation companies. Despite this tendency, a distinctive feature characterising this peasant society today is the persistence of traditional institutions to defend collective landownership. Herrera and Tetoe analyse this connection to land, customary practices and management, and collective ownership of land in a context where land is always the object of desire of private interests and under pressure to be registered and privatised. The authors explain the ambivalence in the position of the state, which faces pressures from foreign investors and international donors, to the point that the dominance of traditional collective forms of social organisation within the unusual structure of land tenure has not prevented the increased export of minerals, hydrocarbons and agribusiness products. The protective role of the state over customary land use has only been effective where private interests are not involved and no natural resources have been discovered. Elsewhere, the state has been taking over land to sell the rights to exploitation of all resources. The access to natural resources and their exploitation by foreign transnationals are being carried out with the support of the state, which articulates this process of land appropriation with the previous ancestral structures of collective landownership, without introducing 'free' land markets.

Even though the logic of 'ideology of landownership' is gaining ground and many peasants have been receptive to financial compensation (e.g. distribution of royalties), the social structures instead of collapsing have adapted to it. Despite constant and convergent pressures towards individualisation of landownership by foreign transnationals, the governments of developed countries as well as international institutions, successive Papua New Guinean authorities have not succeeded in challenging customary collective landownership. The reason is to be found in the legitimate popular resistance by the peasant society against privatisation of land, the imposition of modern register for lands and their management by capitalist laws.

Herrera and Tetoe trace the history of the registration of customary land and the establishment of cadastral systems from the Australian colonial administration to the recent 'land reform' component of the SAPs, that have been jointly imposed by the Papua New Guinean state and foreign donors like Australian Cooperation, USAID, IMF, the World Bank and the Asian Development Bank. They affirm the legitimacy of popular mobilisations gathering large sections of civil society (and even fractions of the military) against privatisation of customary land as common patrimony as well as the legitimacy of their revendication for social progress, in one of the countries with the lowest social indicators in the world.

What is defended is the legitimacy of the principle of collective landowning and free access to the peasant community land; what is demonstrated is the possibility of other rules for land use; and what is recommended is to maintain the existence of non-capitalist peasant farming. Potent constraints obliterate the struggles of a people longing to master their collective destiny. The government has little room for manoeuvre. But an alternative to neoliberalism is required, along with the emergence of a class alliance around the peasantry, to draw a modern development strategy that benefits the Papua New Guinean people.

Chapter 7, co-written by Gérard Choplin and members of the team of European Coordination of La Via Campesina (J. Berthelot, C. Boisgontier, G. Kastler R. Louail, P. Nicholson, J. Riffaud, G. Savigny, J. Verlinden), examines the difficulties of European agriculture, which is very diverse in its productions and structures, as well as the struggles of farmers in this continent. Most of these farmers receive incomes lower than the minimum wages of other professional categories and live under the pressure of repeated sectoral crises due to neoliberal policies and the risk of elimination of their small- or medium-sized farms. While agricultural work is poorly recognised and the environment is threatened, subsidies intended to compensate prices that are often below production costs primarily benefit a minority of large producers and agribusinesses and lead to dumping on the Southern countries. The confrontation is not between North and South, but between two visions of agriculture: agricultural liberalisation and food sovereignty. The authors demonstrate that a Europe without farmers would not provide proof of European development. Things must change and they will change only if European farmers and citizens act together, in solidarity with Southern peasant movements, to draw societies out from their submission to transnational corporations and their logic of maximising private profits.

In the first part, Choplin et al. explain in detail the common problems encountered by farmers, in spite of their diversities, who are dealing with industrialisation and globalisation of production: pressure of productivism, disappearance of small farmers, attacks on peasant agriculture by agribusiness, indebtedness and bankruptcy, outsourcing of agricultural production, monoculture plantations, dissemination of GMOs, pollution, etc. In the face of these destructive tendencies and the inertia of professional organisations defending the interests of dominant economic powers, European farmers have started to resist. The authors describe the evolutions of these struggles, culminating in the emergence of a European farmer movement, connected with civil society and international movements, to propose alternatives, from the European Farmers Coordination to the European Coordination of La Via Campesina, and from local–national to globalised struggles: against the concentration of lands by large farms and agribusiness (by the French farmers from Larzac, for example), the introduction of GMOs (and genetic-transgenic technologies imposed by Monsanto and others), the appropriation of seeds by seed industrial firms, or current neoliberal agricultural policies and rules of international trade promoted by the WTO.

The authors analyse the alternatives opened by the global crisis of the dominant system. According to them, the tasks of the European farmers should be to make food sovereignty (conceived as a right and a duty) the framework of agricultural policies and to build a large alliance of European citizens – producers and consumers – to achieve this goal; to promote a new farming model generating employment, a well-nourished population, and respect for the environment; to work towards global food governance; and to participate in international mobilisations for the defence of nature, climate and biodiversity under attack by WTO free-trade agreements. Grassroots initiatives to relocate food production have multiplied today in the continent.

Finally, the European Coordination of La Via Campesina team concludes that another European common agricultural and food policy is possible, which presupposes deep changes in priorities. The latter should strive to maintain and develop a sustainable and social peasant agriculture, feeding the people, preserving health and the environment and keeping rural landscapes alive; to guarantee peasants decent living conditions thanks to stable and sufficient incomes and greater recognition and improved attractiveness of their profession; to relocate food as much as possible; and in allocating public support, to prioritise productions that are effectively beneficial for employment and the environment.

1

THEORETICAL FRAMEWORK

Food Sovereignty and the Agrarian Question:
Constructing Convergence of Struggles within Diversity

Samir Amin

This first chapter provides a series of analytical elements to answer some of the major questions of our times on agriculture: (1) What kind of agriculture – capitalist, socialist, peasant – can guarantee food sovereignty without which the construction of a multi-polar society is impossible? (2) Which food productions should benefit from top priority in the decision-making process for development? (3) How does one conciliate the growth needed for food production with the preservation of the viability of the earth for the generations to come? The present contribution – in defence of the peasant solution – will put the emphasis on building convergences of the struggles operating in diverse conditions in the North and in the South of our planet.

Family Agriculture in the Present World: Convergences and Differences between the North and the South

In the North: An efficient family agriculture perfectly integrated into dominant capitalism

Modern family agriculture, dominant in Western Europe and in the United States, has clearly shown its superiority over other forms of agricultural production. Annual production per worker (the equivalent of 1,000 to 2,000 tonnes of cereal) has no equal and it has enabled a tiny section of the active population (about 5 per cent) to supply the whole country abundantly and even produce exportable surpluses (Berthelot, 2001). Modern family agriculture has also shown an exceptional capacity for absorbing innovations and much flexibility in adapting to the demand.

This agriculture does not share a specific characteristic of capitalism, that is, its main mode of labour organisation. In the factory, the number of workers enables an advanced division of labour, which is at the origin of the leap in productivity. In the agricultural family business, labour supply is reduced to one or two individuals (the farming couple), sometimes helped by one, two or three associates or permanent labourers, but also, in certain cases, a larger number of seasonal workers, particularly for the harvesting of fruit and vegetables (FAO, 2006). Generally speaking, there is not a definitively fixed division of labour, the tasks being polyvalent and variable. In this sense, family agriculture is not capitalist. However, this modern family agriculture constitutes an inseparable part of the capitalist economy into which it is fully integrated.

In this family agricultural business, self-consumption no longer counts. It depends entirely for its economic legitimacy on its production for the market. Thus the logic that commands the production options is no longer the same as that of the agricultural peasants of yesterday – analysed by Chayanov (1986) – or of today in Third World countries.

The efficiency of the agricultural family business is due to its modern equipment. These businesses possess 90 per cent of the tractors and other agricultural equipment in use in the world (Mazoyer and Roudart, 1997). The machines are 'bought' (often on credit) by the farmers and are therefore their 'property'. In the logic of capitalism, the farmer is both a worker and a capitalist and his income should correspond to the sum of the wages for his work and the profit from his ownership of the capital being used. But it is not so. The net income of farmers in each country is comparable to the average wage earned in industry in that country (UNDP, various years). The state intervention and regulation policies in Europe and the United States, where this form of agriculture dominates, have as their declared objective the aim of ensuring (through subsidies) the equality of 'peasant' and 'worker' incomes (CETIM and GRAIN, 2012). The profits from the capital used by farmers are therefore collected by segments of industrial and financial capital further up the food chain. Control over agricultural production also operates down the food chain by modern commerce (particularly the supermarkets).

In the family agriculture of Europe and the United States, the component of the land rent, which is meant to constitute, in conventional economics, the remuneration of land productivity, does not figure in the remuneration of the farmer–owner, or the owner (when he is not the farmer). The French model of 'anaesthetising the owner' is very telling: in law, the rights of the farmer are given priority over those of the owner. In the United States, where 'respect for property' always has the absolute priority, the same result is obtained by forcing de facto almost all the family businesses to be owners of the land that they farm. The rent of ownership thus disappears from the remuneration of the farmers (Amin, 2005).

The efficiency of this family agriculture is also due to the fact that each unit farms (as owner or not) enough good land: neither too small nor pointlessly large. The area farmed, corresponding, for each stage of the development of mechanised equipment, to what a farmer alone (or a small family unit) can work, has gradually been extended in the interest of efficiency, as Marcel Mazoyer and Laurence Roudart's (1997) analysis of the facts has convincingly demonstrated.

In actual fact, therefore, the agricultural family unit, efficient as it certainly is, is only a subcontractor, caught in the pincers between upstream agribusiness (which imposes selected seeds today, GMOs tomorrow), industry (which supplies the equipment and chemical products), finance (which provides the necessary credits), and downstream in the commercialisation of the supermarkets. The status of the farmer is more like that of the artisan (individual producer) who used to work in the 'putting-out' system (the weaver, for example, being dominated by the merchant that supplied him with the thread and sold the material produced).

It is true that this is not the only form of agriculture in the modern capitalist world. There are also large agribusiness enterprises, that is, big owners who employ many waged labourers (when these estates are not leased out to tenant family farmers). This was generally the case with land in the colonies and still is the case in South Africa (this form of latifundium having been abolished by the agrarian reform of Zimbabwe). There are various forms in Latin America; sometimes they are very 'modernised' (that is, mechanised), as in the Southern Cone of the continent (southern Brazil, Argentina, Uruguay and Chile), and sometimes not. But family agriculture remains dominant in Europe and the United States.

'Really existing socialism' carried out various experiments in 'industrial' forms of agricultural production. The 'Marxism' underlying this option was that of Karl Kautsky who, at the end of the nineteenth century, had 'predicted', not the modernisation of the agricultural family business (its equipment and its specialisation), but its disappearance altogether in favour of large production units, like factories, believed to benefit from the advantages of a thoroughgoing internal division of labour (Kautsky, 1988). This prediction did not materialise in Europe and the United States. However, the myth that it transmitted was believed in the Soviet Union, Eastern Europe (with some nuances), China, Vietnam (in the modalities specific to that country) and, at one time, Cuba. Independently of the other reasons that led to the failure of these experiments (e.g. bureaucratic management, bad macroeconomic planning, reduction of responsibilities due to lack of democracy), there were also errors in judgement about the advantages of the division of labour and specialisation, extrapolated –

without any justification – from certain forms of industry and applied to other fields of production and social activity.

While the reasons for this failure are now recognised, this cannot be said for the forms of capitalist agriculture in the regions of Latin America and Southern Africa mentioned earlier. And yet, the failure is also obvious, despite the profitability and the competitiveness of these modernised forms of latifundia. For this profitability is obtained through horrific ecological wastage (irreversible destruction of productive potential and of arable land) as well as social exploitation (miserable wages).

In the South: Poor peasant cultivators as part of a dominated peripheral capitalism

Peasant agriculture in the South constitutes almost half of humanity – 3 billion human beings. These types of agriculture vary: there are those that have benefited from the green revolution (fertilisers, pesticides and selected seeds) although they are not very mechanised, but their production has risen to between 100 and 500 quintals per labourer; and then there are those which are the same as before the revolution whose production is only around 10 quintals per labourer. The ratio between the average production of a farmer in the North and that of peasant agriculture, which was 10:1 before 1940, is now 100:1. In other words, the rate of progress in agricultural productivity has largely outstripped that in other activities, bringing about a lowering of the real price from 5 to 1 (Mazoyer and Roudart, 1997).

This peasant agriculture in the countries of the South is also well and truly integrated into local and world capitalism. However, closer study immediately reveals both the convergences and the differences in the two types of 'family' economy.

The differences are huge – they are visible and undeniable: the importance of subsistence food in the peasant economies, the only way of survival for those rural populations; the low efficiency of this agriculture, not equipped with tractors or other materials and often highly parcellised; the poverty of the rural world (three-quarters of the victims of undernourishment are rural [Delcourt, 2010]); the growing incapacity of these systems to ensure food supplies for their towns; the sheer immensity of the problems as the peasant economy affects nearly half of humanity.

In spite of these differences, peasant agriculture is already integrated into the dominant global capitalist system. As to the extent of its contribution to the market, peasant agriculture depends on bought inputs and it is the victim of the oligopolies that control the marketing of these products. For the regions having 'benefited' from the 'green revolution' (or half of the peasantry of the South [Mazoyer, 2002]), the siphoning off by dominant capital of profits on the products, both upstream and downstream, is very great. But profits are also siphoned off, in relative terms, for the other half of the peasantry of the South, taking into account the weakness of their production.

Is the modernisation of the agriculture of the South by capitalism possible and desirable?

Let us use the hypothesis of a strategy for the development of agriculture that tries to reproduce systematically in the South the course of modern family agriculture in the North. One could easily imagine that if some 50 million more modern farms were given access to large areas of land for their activities (taking it from the peasant economy and, of course, choosing the best soils) and if they had access to the capital markets, enabling them to equip themselves, they could produce the essentials of what the creditworthy urban consumers still currently obtain from peasant agriculture. But what would happen to the billions of non-competitive peasant producers? They would be inexorably eliminated in a short period of time, i.e. a few decades. What would happen to these billions of human beings, most of whom are already the poorest of the poor, but who feed themselves, for better or for worse – and for a third of them, for worse? No industrial development, more or less competitive, even in the far-fetched hypothesis of a continual yearly growth of 7 per cent for three-quarters of humanity, could absorb even a third of this labour reserve within a period of 50 years. Capitalism, by its nature, cannot resolve the peasant question: the only prospects it can offer are a planet full of slums and 'too many' billions of human beings.

We have therefore reached the point where to open up a new field for the expansion of capital ('the modernisation of agricultural production'), it is necessary to destroy – in human terms – entire societies. Fifty million new efficient producers (200 million human beings with their families) on the one hand, and 3 billion excluded people on the other. The creative aspect of the operation would be merely a drop of water in the ocean of destruction that it would require. I thus conclude that capitalism has entered into its phase of declining senility: the logic of the system is no longer able to ensure the simple survival of humanity (Amin, 1997, 1998). Capitalism is becoming barbaric and leads directly to genocide. It is more than ever necessary to replace it with other development logics that are more rational.

So, what is to be done? It is necessary to accept the continuation of peasant agriculture in the foreseeable future in the twenty-first century. Not due to romantic nostalgia, but quite simply because the solution to the problem is to overtake the logic that drives capitalism and to participate in the long, secular transition into world socialism. It is therefore necessary to work out regulation policies for the relationships between the 'market' and peasant agriculture. At the national and regional levels, these regulations, specific and adapted to local conditions, must protect national production, thus ensuring the indispensable food sovereignty of nations – in other words, delinking the internal prices from those of the so-called global market – as they must do. A gradual increase in the productivity of peasant agriculture, which will doubtless be slow but continuous, would make it possible to control the exodus of the rural populations to the towns. At the level of what is called the global market, the desirable regulation can probably be put in place through inter-regional agreements that meet the requirements of a development that integrates people rather than excludes them.

There is no alternative to food sovereignty

At the global level, food consumption is assured, for 85 per cent of it, by local production (FAO, 2013). Nevertheless, this production corresponds to very different levels of satisfaction of food needs: excellent for North America and West and Central Europe, acceptable in China, mediocre for the rest of Asia and Latin America, and disastrous for Africa. One can also see a strong correlation between the quality and the levels of industrialisation of the various regions: countries and regions that are more industrialised are able to feed their populations well from their own agricultural produce.

The United States and Europe have understood the importance of food sovereignty well and have successfully implemented it through systematic economic policies. But, apparently, what is good for them is not good for the others. The World Bank, the OECD and the European Union try to impose an alternative on the Third World countries, which is 'food security' (for an overview, see FAO, 1983). According to them, these countries do not need food sovereignty and should rely on international trade to cover the deficit in their food requirements, however large it may be. This is perhaps easy for those countries that are large exporters of natural resources (oil, uranium, etc.). For the others, the 'advice' of the Western powers is to specialise their agriculture as much as possible in the production of agricultural commodities for export (cotton, tropical oils, and agro-fuels in the future). The defenders of 'food security' – for others, not for themselves – do not take into account the fact that this specialisation, which has been practised since colonisation, has not made it possible to improve the miserable food rations of the peoples concerned, especially the peasants.

Thus, the advice to peasants who have not yet set foot in the industrial era (e.g. in Africa) is not to engage in 'insane', 'negative' or 'aberrant' industrialisation projects. These are some of the terms used by authors (including experts of the World Bank) who go so far as to attribute the failure of agricultural development in Africa to the industrialisation option of their governments. It is precisely those countries that have taken this 'insane' option (e.g. Korea and China) that have become 'emerging countries' and are able to feed their population better (or less badly), and those that have not done so (in Africa) that are besieged by chronic malnutrition and famine.

This does not appear to embarrass the defenders of the so-called principle of 'food security' – or more accurately, 'food insecurity'. There is little doubt that underneath this obstinacy against Africa committing itself to the path that the success of Asia has inspired lies more than a touch of contempt (if not racism) towards the people. It is regrettable that such condescension is to be found in many Western circles and organisations with good intentions, such as NGOs and even research centres. The complete failure of the 'food security' option is demonstrated by governments that thought they could provide for the needs of their poor urban population through exports (oil among others). They now find themselves trapped by the food deficit that is growing at an alarming rate as a result of these policies. For the other countries, particularly the African ones, the situation is even more disastrous.

On top of this, the economic crisis initiated by the financial collapse of 2008 is already aggravating the situation – and will continue to do so. It is sadly amusing to note how the partners of the OECD (such as the EU institutions) are clinging to the so-called food security policies at a time when the ongoing crisis clearly illustrates their failure. It is not that the governments of the Triad (USA, Europe, Japan) do not 'understand' the problem; this would be to deny them the intelligence that they certainly possess. So can one dismiss the hypothesis that 'food insecurity' is a consciously adopted objective? Has not the 'food weapon' already been deployed? Thus, there is another reason for insisting that without food sovereignty, no political sovereignty is possible. But while there is no alternative to food sovereignty, its efficient implementation does in fact require the commitment to the construction of a diversified economy and hence industrialisation.

The Struggles of the Peasants in the South for the Access to Land

As the access to land depends on 'tenure status', two types of land tenure system must first of all be defined: those that are based on the private ownership of farmland and those that are not.

Land tenure based on the private ownership of land

In this case, the owner has to use the terms of Roman law, *usus* (the right to use an asset), *fructus* (the right to appropriate the returns from the asset) and *abusus* (the right to transfer). This right is 'absolute' in the sense that the owner can farm his land himself, rent it out or even abstain from farming. The property may be given away or sold and it forms part of assets that can be inherited.

Certainly, this right is often less absolute than it appears. In all cases use is subject to public order laws (such as those prohibiting its unlawful use for the cultivation of stupefacients) and, increasingly, to environmental regulations. In some countries where an agrarian reform has been carried through, a limit has been established for the maximum surface area an individual or family can own. The rights of tenant farmers (duration and guarantee of lease and amount of land rent) limit those of the owners in varying degrees to the extent of affording the tenant farmers the major benefit of the protection of the state and its agricultural policies (this is the case in France [Braudel, 1986]). Freedom to choose the crops is not always allowed. In Egypt, the state agricultural services have for a long time determined the proportion of land allotted to different crops depending on their irrigation requirements (Amin, 2011).

This system of landownership is modern inasmuch as it is the product of the constitution of ('really existing') historic capitalism, which first originated in Western Europe (England) and among the Europeans who colonised America. It was established through the destruction of the 'customary' systems for regulating access to land, even in Europe. The statutes of feudal Europe were based on the superposition of rights to the same land: those of the peasant concerned and other members of a village community (serfs or freemen), those of the feudal lord, and those of the king. The assault on these rights took the form of 'enclosures' in England, imitated in different ways in all European countries during the course of the nineteenth century. Very early on in Volume 1 of *Capital*, Marx (1976) denounced this radical transformation, which excluded the majority of the peasants from access to use of the land, turning them into proletariat emigrants to the towns (forced by circumstance). He regarded the case of those who stayed on as farm labourers or tenant farmers as among the measures of primitive accumulation that dispossessed the producers both of property and of the use of the means of production.

The use of the terms of Roman law (*usus* and *abusus*) to describe the status of modern bourgeois ownership perhaps indicates that the latter had distant 'roots' – in this case, in landownership in the Roman Empire and, more precisely, in pro-slavery latifundist ownership. The fact remains that as these particular forms of ownership have disappeared in feudal Europe, we cannot talk of the 'continuity' of a 'western' concept of ownership (itself associated with 'individualism' and of the values it represents), which has, in fact, never existed.

The rhetoric of capitalist discourse about its 'liberal' ideology has produced not only this myth of 'western continuity', but, above all, another even more dangerous myth, namely that of the absolute and superior rationale of economic management based on the private and exclusive ownership of the means of production, which it considers farmland to be. In fact, according to conventional economics the 'market', that is, the transferability of ownership of capital and land, determines the optimal (most efficient) use of these 'factors of production'. So, according to this principle, land becomes a 'merchandise like any other', transferable at the 'market' price, in order to guarantee that the best use is made of it both for the owner and society as a whole. This is mere tautology, yet the whole ('vulgar', or to use Marx's term, 'acritical') bourgeois economic discourse is based on it.

This same rhetoric is used to legitimise the principle of landownership by dint of the fact that it alone can guarantee that the farmer who invests to improve his yield per hectare and the productivity of his work (and that of any employees) will not suddenly be dispossessed of the fruit of his labour and savings. This is not the case, and other forms of regulating the right to use the land can produce similar results. To sum up, this dominant discourse draws the conclusions that it sees fit from the construction of western modernity in order to propose them as the only necessary 'rules' for the advancement of all other peoples. To make land everywhere private property in the current sense of the term, as practised in capitalist centres, is to spread the policy of 'enclosures' all over the world, in other words, it is to hasten the dispossession of the peasants.

This course of action is not new; it began and continued through the earlier centuries of the global expansion of capitalism in the context of colonial systems in particular. Today the WTO intends only to accelerate the process even though the destruction that would result from this capitalist approach is becoming increasingly foreseeable and predictable. Resistance to this option by the peasants and peoples affected would make it possible to build a real and genuinely human alternative.

Land tenure systems not based on the private ownership of land

As we can see, this definition is in the negative – *not* based on private property – and therefore cannot be designated to a homogeneous group since access to land is regulated in all human societies. However, it is regulated either by 'customary authorities', 'modern authorities', the state or, more specifically, and more often, by a group of institutions and practices involving individuals, communities and the state.

'Customary' administration, expressed in terms of customary law or known as such, has always or almost always ruled out private property in the modern sense, and has always guaranteed access to land to all families (rather than individuals) concerned – in other words, to those that are part of a 'village community' that is distinct and can be identified as such. Yet it has (almost) never guaranteed 'equal' right to land. In the first place, it most often excluded 'foreigners' (usually the vestiges of conquered peoples) and 'slaves' (of differing status) and shared land unequally depending on clan membership, lineage, caste or status ('chiefs', 'free men', etc.). So there is no reason to heap excessive praise upon these traditional rights, as a number of anti-imperialist national ideologues unfortunately do. Progress will certainly require them to be challenged.

Customary administration has almost never been the system used in 'independent villages'. These have always been part of stable or changing, sound or precarious state groupings depending on circumstances, but very rarely have they been absent. So, the rights of use of the communities and families that made them up have always been limited by those of the state, which levied taxes. This is why I describe the vast family of pre-modern production methods as 'tributary' (Amin, 1977, 1978, 1980).

These complex forms of 'customary' administration, which differ from one time and place to another, only persist, in the best of cases, in extremely deteriorated forms and have been under attack by the dominant rationale of world capitalism for at least two centuries (in Asia and Africa), and sometimes five (in Latin America).

In this respect, India is probably one of the clearest examples. Before British colonisation, access to land was managed by 'village communities', or more precisely by their ruling upper castes and classes. The lower castes or the Dalits were excluded from this and were treated as a kind of collective slave class similar to the Hilotes of Sparta. These village communities were, in turn, controlled and exploited by the imperial Mughal state and its vassals (states of the Rajahs and other rulers), which levied tribute. The British raised the status of the zamindars, formerly land revenue collectors, to that of 'owners'. These zamindars became large allied landowners in spite of tradition – although they upheld 'tradition' when it suited them, for example, by 'respecting' the exclusion of Dalits from access to land. Independent India has not challenged this serious colonial inheritance, which has been the cause of the incredible poverty of the majority of its peasantry and, later, of its urban proletariat. The solution to these problems and the building of a viable economy for the peasant majority is possible through an agrarian reform in the strictest sense of the term. The European and US colonisations of South-East Asia and the Philippines respectively resulted in similar developments. The 'enlightened despotic' regimes of the East (the Ottoman Empire, the Egypt of Mohamed Ali and the Shahs of Iran) also by and large established private ownership in the modern sense to the benefit of a new class, wrongly described as 'feudal' (by most historical Marxist thinking) and recruited from among the senior ranks of their power system.

As a result, since then, private ownership of land has affected the majority of farmland, especially the best, throughout Asia, excluding China, Vietnam and the former Soviet republics of Central Asia. There are only remnants of deteriorated para-customary systems in the poorest regions that are of the least value to the dominant capitalist farming in particular. These structures differ widely, juxtaposing large landowners (country capitalists), rich peasants, middle peasants, poor peasants and the landless. There is no peasant 'organisation' or 'movement' that transcends these acute class conflicts.

In Arab Africa, South Africa, Zimbabwe and Kenya, the colonisers (with the exception of Egypt) granted their colonists (or the Boers in South Africa) 'modern' private properties of a generally latifundist type. This legacy has certainly been brought to an end in Algeria, but the peasantry there had almost disappeared, proletarianised (and reduced to vagrancy) by the extension of colonial lands, whereas in Morocco and Tunisia the local bourgeoisie took them over (which was also the case to some extent in Kenya). In Zimbabwe, the revolution has challenged the legacy of colonialisation to the benefit, in part, of new middle owners of urban rather than rural origin and, in part, of 'poor peasant communities'. South Africa still remains outside this movement. The remnants of deteriorated para-customary systems that survive in the 'poor' regions of Morocco or Berber Algeria and the former Bantustans of South Africa are threatened with private appropriation from inside and outside the societies concerned.

In all these situations, a scrutiny of the peasant struggles (and possibly those of the organisations that support them) is required: are we talking about 'rich peasant' movements and demands in conflict with some orientation of the state policy (and the influences of the dominant world system on them), or of poor and landless peasants? Can they form an 'alliance' against the dominant (so-called 'neoliberal') system? Under what conditions? To what extent? Can the demands – expressed or otherwise – of poor and landless peasants be 'forgotten'?

In inter-tropical Africa, the apparent survival of 'customary' systems is certainly more visible because here the model of colonisation took a different and unique direction, known in French (the term has no translation in English) as 'économie de traite'. The administration of access to land was left to the so-called 'customary' authorities, though controlled by the colonial state (through traditional clan leaders, legitimate or otherwise, created by the administration). The purpose of this control was to force peasants to produce a quota of specific products for export (peanuts, cotton, coffee, cocoa) over and above what they required for their own subsistence. Maintaining a system of land tenure that did not rely on private property suited colonisation since no land rent entered into determining the prices of the designated products. This resulted in land being wasted, destroyed by the expansion of crops, sometimes permanently, as illustrated by the desertification of peanut producing areas of Senegal.

Yet again capitalism showed that its 'short-term rationale', an integral part of its dominant rationale, was in fact the cause of an ecological disaster. The combination of subsistence farming and the production of goods for export also meant that the peasants were paid almost nothing for their work. To talk in these circumstances of a 'customary land tenure system' is going too far. It is a new regime that preserves only the appearance of 'traditions' and often the least valuable ones.

China and Vietnam

In Asia, China and Vietnam provide unique examples of a land-access administration system that is based neither on private ownership nor on 'customs', but on a new revolutionary right unknown elsewhere – with the exception of Cuba. It is the right of all peasants (defined as inhabitants of a village) to equal access to land, and I stress the use of 'equal'. This right is the finest accomplishment of the Chinese and Vietnamese revolutions.

In China, and even more so in Vietnam, which was more extensively colonised, 'former' land tenure systems (those that I have described as 'tributary' in Amin, 1977, 1978, 1980) were already quite eroded by dominant capitalism. The former ruling classes of the imperial power system had turned most of the agricultural land into private or quasi-private property, whereas the development of capitalism encouraged the formation of new rich peasant classes. Mao Zedong is the first and without doubt the only one, followed by the Chinese and Vietnamese communists, to have defined a revolutionary agrarian strategy based on the mobilisation of the majority of poor, landless and middle peasants. From the outset, the triumph of this revolution made it possible to abolish private ownership of land, which was replaced by state ownership, and organise new forms of equal access to land for all peasants. This organisation has certainly passed through several successive phases, including that inspired by the Soviet model based on production cooperatives. The limited achievements made by the latter have led both countries to return to peasant family farming.

Is this model viable? Can it lead to a sustained improvement in production without bringing about an excess of rural manpower? Under what conditions? What supporting policies does it require from the state? What types of political management can meet the challenge?

Ideally, the model involves the dual affirmation of the rights of the state – sole owner – and of the usufructuary – the peasant family. It guarantees equal distribution of village land among all families and prohibits any use of it other than for family farming, such as renting. It makes sure that the proceeds of investments made by the usufructuary return to him/her in the short term through his/her right of ownership of all farm produce – which is freely marketed, although the state ensures a minimum price – and in the long term by enabling inheritance of usufruct exclusively to the children remaining on the farm (any person who emigrates from the village loses his/her right of access to the land, which is then redistributed). As this involves rich land but also small (even tiny) farms, the system is only viable as long as the vertical investment (the green revolution with no large-scale industrialisation) is as efficient in allowing the increase of production per rural worker as horizontal investment (the expansion of farming supported by increased industrialisation).

Has this 'ideal' model ever been implemented? Certainly something close to it has been, for example, during Deng Xiaoping's time in China. However, the fact remains that although this model ensures a high degree of equality within the village, it has never been able to overcome the inequalities between one community and another that are a function of the quality of the land, the density of the population and the proximity of urban markets. Furthermore, no redistribution system has been up to the challenge, even through the structures of cooperatives and state trade monopolies of the 'Soviet' phase.

Definitely more serious is the fact that the system is itself subject to internal and external pressures, which undermine its direction and social scale. Access to credit and satisfactory subsidisation are subject to bargaining and interventions of all kinds, legitimate or otherwise. Equal access to land is not synonymous with equal access to the best production conditions. The popularisation of the 'market' ideology contributes to this destabilisation. The system tolerates (and has even re-legitimised) farm tenancy and the employment of waged employees. Right-wing discourse – encouraged from abroad – stresses the need to give the peasants in question 'ownership' of the land and to open up the 'farmland market'. It is quite clear that rich peasants (and even agribusiness) seeking to increase their property support this discourse.

This system of peasant access to land has been administered thus far by the state and the party, which are one. Clearly, one might have thought that it could have been administered by genuinely elected village councils. This is certainly necessary, as there is hardly any other means of winning the support of the majority and reducing the intrigues of the minority would-be beneficiaries of a more markedly capitalist approach. The 'party dictatorship' has shown itself to be largely inclined to careerism, opportunism and even corruption. Social struggles are currently far from non-existent in rural China and Vietnam. They are no less strongly expressed than elsewhere in the world but they are by and large 'defensive' and concerned with defending the legacy of the revolution – equal right to land for all. This legacy must be defended, especially as it is under greater threat than might be thought, despite repeated affirmations from both governments that 'state ownership of the land will never be abolished in favour of private property'. Yet, today this defence demands recognition of an equal right to land for all through the organisation of those who are affected, that is, the peasants (Amin, 2013b).

Agrarian reforms and forms of organisation of agricultural production and land tenure

In Asia and Africa, the forms of organisation of agricultural production and land tenure are too varied for one single formula of 'alternative peasant social construction' to be recommended for all.

By 'agrarian reform' we must understand the redistribution of private property when it is deemed too unequally divided. It is not a matter of 'reforming the land tenure status', since we are dealing with a land tenure system governed by the principle of ownership. However, this reform seeks to meet the perfectly legitimate demand of poor and landless peasants and to reduce the political and social power of large landowners. Yet, where it has been implemented – in Asia and Africa after the liberation from former forms of imperialist and colonial domination – this has been done by non-revolutionary hegemonic social blocks, in the sense that they were not directed by the dominated poor classes, who are a majority. The exceptions to this are China and Vietnam, where, in fact, for this reason there has been no 'agrarian reform' in the strict sense of the term, but, as I have already said, suppression of the private ownership of land, affirmation of state ownership and implementation of the principle of equal access to the use of the land by all peasants. Elsewhere, real reforms dispossessed the only large owners to the eventual benefit of middle and even rich peasants (in the longer term), ignoring the interests of the poor and landless. This has been the case in Egypt and other Arab countries. The reform underway in Zimbabwe may have a similar perspective. In other situations, such as in India, South-East Asia, South Africa and Kenya, reform is still on the agenda of what is needed.

Even where agrarian reform is an immediate unavoidable demand, its long-term success is uncertain, as it reinforces an attachment to 'small ownership', which becomes an obstacle to challenging the land tenure system based on private ownership.

Russian history illustrates this tragic situation. The evolution begun after the abolition of serfdom (in 1861), accelerated by the revolution of 1905, and then the policies of Stolypin, had already produced a 'demand for ownership' that the revolution of 1917 consecrated by means of a radical agrarian reform. As we know, the new small owners were not happy about giving up their rights to the benefit of the unfortunate cooperatives created in the 1930s. A 'different approach' based on peasant family economy and generalised small ownership might have been possible, but it was not tried.

Yet, what about the regions (other than China and Vietnam) in which the land tenure system is not (yet) based on private property? We are, of course, talking about inter-tropical Africa.

We return here to an old debate. In the late nineteenth century, Marx, in his correspondence with the Russian Narodniks – Vera Zasulich among others

(Marx, 1881) dares to state that the absence of private property may be a major advantage for the socialist revolution by allowing the transition from a system for administering access to land other than that governed by private ownership (but he does not say what forms this new system should take, and the use of 'collective', however fair, remains insufficient). Twenty years later, Lenin claimed that this possibility no longer existed and had been destroyed by the penetration of capitalism and the spirit of private ownership that accompanied it (see Lenin, 1965).[1] Was this judgement right or wrong? I cannot comment on this matter as it goes beyond my knowledge of Russia. However, the fact remains that Lenin did not consider this issue of crucial importance, having accepted Kautsky's point of view in 'On the Agrarian Question'. Kautsky generalised the scope of the modern European capitalist model and felt that the peasantry was destined to 'disappear' due to the expansion of capitalism itself (see Kautsky, 1988).

In other words, capitalism would have been capable of 'resolving the agrarian question'. Although 80 per cent true for the capitalist centres (that is, the Triad, which is 15 per cent of the world's population), this proposition does not hold true for the 'rest of the world' (that is, 85 per cent of its population). History shows that not only has capitalism not resolved this question for 85 per cent of the people, but from the perspective of its continued expansion, it cannot resolve it any longer (other than by genocide – a fine solution!). So it fell upon Mao Zedong and the communist parties of China and Vietnam to find a suitable solution to the challenge (Amin, 2013a).

The question resurfaced during the 1960s with African independence. The states and party-states that arose from the national liberation movements of the continent enjoyed, in varying degrees, the support of the peasant majority of their peoples. Their propensity to populism led them to conceive of a 'specific (African) socialist approach'. This approach could certainly be described as very moderately radical in its relationship with both dominant imperialism and the local classes associated with its expansion. It did not raise to any lesser extent the question of the rebuilding of peasant society in a humanist and universalist spirit– a spirit that often proved highly critical of the 'traditions' that the foreign masters had in fact tried to use to their advantage.

All, or almost all, African countries adopted the same principle, formulated as an 'inalienable right of state ownership' of all land. I do not believe this proclamation to have been a 'mistake', nor do I think that it was motivated by extreme 'statism'.

Examination of the way that the current peasant system really operates and its integration into the capitalist world economy reveals the scale of the challenge. This management is provided by a complex system that is based on 'custom', private ownership (capitalist) and the rights of the state. The 'custom' in question has degenerated and barely serves to disguise the discourse of bloodthirsty dictators who pay lip service to 'authenticity', which is nothing but a fig leaf that they think hides their hunger for pillage and treachery in the face of imperialism. The only major obstacle to the expansionist tendency of private ownership is the possible resistance of its victims. In some regions that are better able to yield rich crops (irrigated areas and market-garden farms) land is bought, sold and rented with no formal land title.

Inalienable state property, which I defend in principle, itself becomes a vehicle for private ownership. Thus, the state can 'provide' the land needed for the development of a tourist area, a local or foreign agribusiness or even a state farm. The land titles necessary for access to better areas are distributed in a way that is rarely transparent. In all cases the peasant families who inhabited the areas and are asked to leave are victims of these practices, which are an abuse of power. Still, the 'abolition' of inalienable state property in order to transfer it to the occupiers is not feasible in reality (all village lands would have to be registered with the land registry), and if this were attempted it would only allow rural and urban notables to help themselves to the best plots.

The right answer to the challenges of the management of a land tenure system not based on private ownership (as the main system at least) is through state reform and its active involvement in the implementation of a modernised and economically viable and democratic system for administering access to land that rules out, or at least minimises, inequality. The solution certainly does not lie in a 'return to customs', which would, in fact, be impossible and only serve to accentuate inequalities and open the way for savage capitalism.

We cannot say that no African state has ever tried the approach recommended here. Following Mali's independence in September 1961, the Sudanese Union began what has very wrongly been described as 'collectivisation'. In fact, the cooperatives that were set up were not productive cooperatives, and production remained the exclusive responsibility of family farms. It was a form of modernised collective authority that replaced the so-called 'custom' on which colonial authority had depended. The party that took over this new modern power was clearly aware of the challenge and set the objective of abolishing customary forms of power that were deemed to be 'reactionary', even 'feudal'. It is true that this new peasant authority, which was formally democratic (those in charge were elected), was in actual fact only as democratic as the state and the party.

However, it had 'modern' responsibilities, namely, to ensure that access to land was administered 'correctly', that is, without 'discrimination', to manage loans, the distribution of subsidies (supplied by state trade), and product marketing (also partly the responsibility of state trade). In practice, nepotism and extortion have certainly never been stamped out. The only response to these abuses should have been the progressive democratisation of the state and not its 'retreat', as liberalism then imposed (by means of an extremely violent military dictatorship) to the benefit of the traders (dioulas).

Other experiences in the liberated areas of Guinea Bissau impelled by theories put forward by Amilcar Cabral and in Burkina Faso at the time of Sankara have also tackled these challenges head on and sometimes produced unquestionable progress that people try to erase today. The creation of elected rural collectives in Senegal is a response whose principle I would not hesitate to defend. Democracy is a never-ending process, no more so in Europe than in Africa.

Alternatives

What current dominant discourse understands by 'reform of the land tenure system' is quite the opposite of what the construction of a real alternative based on a prosperous peasant economy requires. This discourse, promoted by the propaganda instruments of collective imperialism – the World Bank, numerous cooperation agencies, as well as a number of NGOs with considerable financial backing – understands land reform to mean the acceleration of privatisation of land and nothing more. The aim is clear: create conditions that allow 'modern' islands of foreign or local agribusiness to take possession of the land they need in order to expand. Yet, the additional produce that these islands could provide (for export or creditworthy local market) will never meet the challenge of the requirements of creating a prosperous society for all, which implies the advancement of the peasant family economy as a whole.

So, counter to this, a land tenure reform conceived from the perspective of the creation of a real, efficient and democratic alternative supported by prosperous peasant family production must define the role of the state (principal inalienable owner) and the institutions and mechanisms of administering access to land and the means of production.

I do not exclude here complex mixed formulas that are specific to each country. Private ownership of land may be acceptable – at least where it is established and held to be legitimate. Its redistribution can or should be reviewed, where necessary, as part of an agrarian reform (South Africa, Zimbabwe and Kenya, with respect to Sub-Saharan Africa). I would not necessarily even rule out the controlled clearance of land for agribusiness in all cases. The key lies elsewhere, in the modernisation of peasant family farming and the democratisation of the management of its integration into the national and global economy. This is no blueprint to propose for these areas, so I will limit myself to pointing out some of the great problems that this reform poses.

The democratic question is indisputably central to the response to the challenge. It is a complex and difficult question that cannot be reduced to an insipid discourse about good governance and electoral pluralism. There is an undeniable cultural aspect to the question: democracy leads to the abolition of 'customs' that are hostile to it (prejudice concerning social hierarchies and, above all, the treatment of women). There are legal and institutional aspects to be considered: creating systems of administrative, commercial and personal rights that are consistent with the aims of the plans for social construction and establishment of suitable (generally elected) institutions. However, above all, the progress of democracy will depend definitively on the social power of its defenders. The organisation of peasant movements is, in this respect, absolutely irreplaceable. It is only to the extent that peasants are able to express themselves that progress will be made in the direction known as 'participative democracy' (as opposed to the reduction of the problem to the dimension of 'representative democracy').

Relations between men and women are another aspect of the democratic challenge that is no less essential. Peasant 'family farming' obviously concerns the family, which is to this day characterised almost everywhere by structures that require the submission of women and the exploitation of their workforce. Democratic transformation will not be possible in these conditions without the organised action of the women concerned.

Attention must be given to the question of migration. In general, 'customary' rights exclude 'foreigners' (that is, all those who do not belong to the clans, lineages and families that make up the village community in question) from the right to land or place conditions upon their access to it. Migration resulting from colonial and postcolonial development has sometimes been at such a large scale that it has overturned the concepts of ethnic 'homogeneity' in the regions affected by this development. Immigrants from outside the state in question (such as the Burkinabe in Ivory Coast) or those who are formally citizens of the state but of an 'ethnic' origin other than the regions they have made their homes (like the Hausa in the Nigerian state of Plateau), see their rights to the land they have cultivated challenged by short-sighted and chauvinistic political movements that also have foreign support. To throw the 'communitarianism' in question into ideological and political disarray and uncompromisingly denounce the paracultural discourse that underpins it has become one of the indispensable conditions of real democratic progress.

The analyses and propositions set out here only concern the status of tenure or rules on access to land. These matters are certainly central to debates on the future of agricultural and food production, peasant societies and the people that make them up, yet they do not cover all aspects of the challenge. Access to land remains devoid of the potential to transform society if the peasants who benefit from it cannot have access to the essential means of production under suitable conditions (credit, seed, subsidies, access to markets). Both national policies and international negotiations that aim to define the context in which prices and revenues are determined are other aspects of the peasant question.

Further information on these questions that go beyond the scope of the subject we are dealing with here can be found in the writings of Jacques Berthelot – a critical analyst of projects to integrate agricultural and food production into 'global' markets (see in particular Berthelot, 2001). Therefore, we shall restrict ourselves to the two main conclusions and proposals reached:

1. We cannot allow agricultural and food production, and land, to be treated as ordinary 'merchandise' and then agree to the need to integrate them into plans for global liberalisation promoted by the dominant powers (the United States and Europe) and transnationalised capital.

    The agenda of the WTO, which inherited the General Agreement on Tariffs and Trade (GATT) in 1995, must quite simply be refused. In Asia and Africa, peasant organisations, social and political forces that defend the interests of popular classes and the nation (and demands for food sovereignty in particular), as well as those who have not given up on a development project worthy of its name, must be persuaded that negotiations entered into as part of the WTO agenda can only result in catastrophe for the peoples of Asia and Africa. It must be emphasised that this agenda threatens to devastate the lives of more than 2.5 billion peasants from the two continents while offering them no other prospect than migration to slums, being shut away in 'concentration camps', the construction of which is already planned for the unfortunate future emigrants (Amin, 2008).

    Capitalism has reached a stage where its continued expansion requires the implementation of 'enclosure' policies on a global scale, like the enclosures at the beginning of its development in England – except that today, the destruction of the 'peasant reserves' of cheap labour on a global scale will be nothing less than the genocide of half of humanity. On the one hand is the destruction of the peasant societies of Asia and Africa; and on the other, some billions in extra profit for global capital and its local associates, derived from a socially useless production, since it is not destined to satisfy the unsolvable needs of hundreds of millions of extra hungry, but to increase the number of obese in the North and those who emulate them in the South.

    So Asian and African states must quite simply be called upon to withdraw from these negotiations and therefore reject decisions taken by the imperialist United States and Europe within the famous 'Green Rooms' of the WTO. This voice must be made to be heard and the governments concerned must be forced to ensure that it is heard in the WTO.

2. We can no longer accept the behaviour of the major imperialist powers (the United States and Europe) that together assault the people of the South within the WTO. It must be pointed out that the same powers that try to unilaterally impose their 'liberalist' proposals on the countries of the South

do not abide by these proposals themselves and behave in a way that can only be described as systematic cheating.

The Farm Bill in the United States and the agricultural policy of the European Union violate the very principles that the WTO is trying to impose on others. The 'partnership' projects proposed by the European Union following the Cotonou Convention as of 2008 are really 'criminal', to use the strong but fair expression of J. Berthelot (2012).

So we can and must hold these powers to account through the authorities of the WTO set up for this purpose. A group of countries from the South not only could but also must do it.

Asian and African peasants organised themselves in the previous period of their peoples' liberation struggles. They found their places in powerful historical blocks that enabled them to be victorious over the imperialism of the time. These blocks were sometimes revolutionary (China and Vietnam) and found their main support in rural areas among the majority classes of middle, poor and landless peasants. When, elsewhere, they were led by the national bourgeoisie, or those among the rich and middle peasants who aspired to becoming bourgeois, large landowners and 'customary' local authorities in the pay of colonisation were isolated.

Having turned over a new leaf, the challenge of the new collective imperialism of the Triad will only be lifted if historical blocks form in Asia and Africa that cannot be a remake of the former ones. The definition of the nature of these blocks, their strategies and their immediate and longer-term objectives in these new circumstances, is the challenge facing the alter-globalist movement and its constituent parts in social forums. This is a far more serious challenge than is imagined by a large number of movements engaged in current struggles.

New peasant organisations exist in Asia and Africa that support the current visible struggles. Often, when political systems make it impossible for formal organisations to form, social struggles for the campaign take the form of 'movements' with no apparent direction. Where they do exist, these actions and programmes must be more closely examined: What peasant social forces do they represent? Whose interests do they defend? – those of the majority mass of peasants or those of the minorities that aspire to find their place in the expansion of dominant global capitalism?

We should be wary of instantaneous replies to these complex and difficult questions. We should not 'condemn' organisations and movements for not having the support of the majority of peasants for their radical programmes. That would be to ignore the demands of the formation of large alliances and strategies in stages. Neither should we subscribe to the discourse of 'naive alter-globalism' that often sets the tone of forums and fuels the illusion that the world would be set on the right track only by the existence of social movements. A discourse, it is true, that is more that of numerous NGOs – well-meaning perhaps – than of peasant and worker organisations.

The analyses and proposals made in this study are only relevant for Asia and Africa. The agrarian questions in Latin America and the Caribbean have their own particular and sometimes unique particularities. Thus, in the Southern Cone, the modernised, mechanised latifundium that benefits from cheap labour is the method of farming that is best adapted to the demands of a liberal global capitalist system that is even more competitive than the agriculture of the United States and Europe.

Note

1. See especially 'The Proletariat and the Peasantry' and 'The "Peasant Reform" and the Proletarian–Peasant Revolution'.

2

LATIN AMERICA

Reflections on the Tendencies of Capital
in Agriculture and Challenges for Peasant
Movements in Latin America

João Pedro Stedile

The aim of this chapter is to briefly present information to foment debate and reflection on the main forms of capital activity in agriculture, and in particular through transnational corporations. There is a natural logic to how capitalism operates in agriculture, now in a phase dominated by financial capital. There are specific characteristics determined by the recent crisis of financial capital that have consequences for the organisation of agricultural production and the life of peasants. The chapter also highlights contradictions that need to be understood in order to act upon them. For one thing, it presents what could be the main elements of a peasant programme for agriculture, especially for the countries of the South, where the peasant way of living in the countryside predominates, and where they suffer more under the power wielded by international capital agricultural technology, production and trade. The chapter also presents some organisational and political challenges for peasant movements at local and international levels because of the current disadvantageous power correlation, where international capital is on the offensive to control nature, production and agricultural goods. This analysis results from the experienced reality in Latin America, especially in Brazil, as a result of the control of agriculture by large capital, and from struggle and resistance by peasant movements and their reflections on how to face capital with an alternative, popular and peasant development model.

Capital Trends in Agriculture

Capital mobility in its current hegemonised phase by international financial capital

The development of the capitalist mode of production has gone through several phases. It started in the fifteenth century as mercantile capitalism and then evolved into industrial capitalism in the eighteenth and nineteenth centuries. In the twentieth century, it developed as monopoly and imperialist capitalism. For the last two decades we are experiencing a new phase of capitalism that is dominated by globalised financial capital, which means that capital accumulation or wealth is concentrated primarily in the sphere of financial capital. This financial capital needs to control the production of goods (in industry, mining and agriculture) and trade around the world to seize the surplus value generated by agricultural workers in general (Carcanholo, 2014).

Internationalised financial capital took control of agriculture through various mechanisms. The first of these was financial surplus capital. Banks began to buy stocks from hundreds of medium and large-scale companies operating in different sectors related to agriculture. Through controlling most of the stocks, they promoted a process of concentration of the companies working on agriculture. In a few years, these companies had achieved an astronomical growth of capital through investments made by financial capital (Vitali, Glattfelder and Battison, 2011). They moved on to control many different sectors related to agriculture, such as trade, production of inputs, agricultural machinery, agro-industries, pharmaceuticals, agrochemicals and tools. It is important to understand that this capital was accumulated outside of agriculture but was applied within it and quickly accelerated the process of growth and concentration, which by normal means of wealth accumulation for agricultural goods would have taken many more years (Herrera, Dierckxsens and Nakatani, 2014).

The second mechanism of control was the process of dollarisation of the global economy. This allowed companies to take advantage of favourable exchange rates to enter national economies and easily buy up companies and take control of the production markets and trade of agricultural goods (Nakatani and Herrera, 2010, 2013).

The third mechanism was the free trade rules imposed by international organisations such as the World Trade Organization (WTO), the World Bank and the International Monetary Fund, as well as multilateral agreements, which regulated the trade of agricultural goods in accordance with the interests of large companies, forcing subservient governments to liberalise this trade. Thus, transnational corporations were able to enter countries and control their national markets for agricultural goods and inputs in virtually the whole world (Berthelot, 2001).

In practically every country, the development of agricultural production has been increasingly dependent on industrial inputs and has been put at the mercy of the use of credit to finance production. These loans facilitate the funding of the offensive of this mode of production in 'industrial agriculture' and the companies that produce inputs. In other words, banks finance the implementation and control of industrial agriculture worldwide.

And finally, in most countries, governments have abandoned the public policies protecting the national agricultural market and the peasant economy (Nicholson, Montagut and Rulli, 2012). They liberalise markets and implement neoliberal subsidy policies only for large capitalist agricultural production. These government subsidies are mainly tax exemptions on exports or imports and implementation of favourable interest rates for capitalist agriculture.

As a result of two decades of the logic of financial capital control on agricultural production, there are now approximately 50 major corporations that control most of the world agricultural production and trade (CETIM and GRAIN, 2012).

The recent crisis of financial capital and its consequences for agriculture and nature's goods

During the years 1990–2008, there was an offensive of financial capital in agriculture (Stedile, 2007b), which in recent years has been intensified by the recurrent crises of financial capital in the United States and Europe. This crisis of financial capital is further aggravating the effects of the control of international capital on peripheral economies, that is, on agriculture and the peasant economy. This has been happening for several reasons.

Large economic groups from the North, due to the crisis in their own countries – low interest rates, the instability of the dollar and their currencies – have fled the North to peripheral economies trying to protect their volatile capital, and have invested in fixed assets such as land, minerals, agricultural raw materials, water, high biodiversity territories, productive investments and agricultural production, as well as in the control of renewable energy sources, such as hydroelectric power or ethanol mills (Transnational Institute, 2007).

The crises of oil prices and their consequences on global warming and the environment has led the automobile–oil complex to start investing large sums of capital in the production of agro-fuels, especially in the production of sugar cane and maize for ethanol, and soybean, peanut, rapeseed and oil palm (African palm) for vegetable oil. This has resulted in an unmitigated attack by financial capital and transnational companies on the Southern tropical agriculture (GRAIN, 2007).

Finally, there is the crisis in which this financial capital has entered the futures agricultural and mining markets to invest its assets and speculate in the futures market or simply transform money into futures goods. This movement has generated a steep rise in the prices of agricultural goods traded by companies in the world futures stock exchange markets (FAO et al., 2011). The average international prices for agricultural goods are no longer related to the average production cost and the actual value measured by the socially necessary labour time, but are rather the result of speculation and oligopolistic control of agricultural markets by these large companies.

The current situation of the transnational corporation and financial capital control over agriculture

There are many aspects one could analyse of the situation and consequence of the action of transnational corporations on agriculture. Here, we will consider only the economic aspects. A few transnational corporations have consolidated the production and world trade of agricultural goods, especially standardised goods like crops and dairy, and now exercise worldwide control over them (CETIM and GRAIN, 2012). They also control the whole production chain of inputs and machines used in agriculture.

An accelerated process of capital centralisation has meant that the same company can now control the production and trade of a range of products and industries (UNCTAD, 2008), such as the manufacture of agricultural inputs (chemical fertilisers, poisons, pesticides), agricultural machinery, pharmaceuticals and GM seeds, as well as of a wide range of products arising from the agro-industry, like food or cosmetic and superfluous goods.

Interdependence among industrial, commercial and financial capital within a company has grown. Now there is an almost absolute control over the prices of agricultural goods and agricultural inputs worldwide. Although prices should have their basis in real value (average labour time), the oligopolistic control of goods generates practices that price goods above their real value, and therefore companies obtain extraordinary profits. At the same time, this leads to the bankruptcy of small and medium companies that cannot produce at the same scale as international corporations (Berthelot, 2008).

A company hegemony has taken hold of scientific knowledge, research (which requires increasingly greater resources) and technologies applied to agriculture, imposing a technological model of so-called 'industrial farming' worldwide, dependent on inputs produced outside of agriculture. This model is presented as the only, the best and the cheapest way for agricultural production, ignoring ancient techniques available in popular knowledge and agroecology. This company hegemony is a consequence of the lack of state investment in agriculture and husbandry research. Throughout the twentieth century, many national states invested public resources in agricultural research and the results were democratised and made accessible to all farmers in each country. Now agricultural knowledge and research have become privatised and the results are used as commodities in order to obtain higher returns (Delcourt, 2010). Often companies charge farmers for using new technologies by embedding royalties in the high market prices of genetically modified seeds or agricultural machinery and pesticides.

Corporate private ownership is imposed on goods available in nature, in particular on genetically modified seeds, and more recently on sources of drinking water and reservoirs for power generation or irrigation. An offensive is also under way in the South attempting the privatisation of territories with a wealth of plant and animal biodiversity (CETIM and GRAIN, 2012).

Excessive concentration exists in the production of agricultural goods, especially those intended for foreign markets, by an ever smaller number of large landowners allied to corporations. The case of Brazil is illustrative of this: about 10 per cent of all agricultural dwellings in the country control 80 per cent of the production value (Stedile, 2002).

These developments are on course for a dangerous standardisation of human and animal foods all over the world. Humanity is being misled into eating more and more food standardised by companies. Food has become a mere commodity that must be consumed, massively and fast. This has incalculable consequences, such as the destruction of local food habits, culture and high risks to human and animal health.

Throughout the world, there is a generalised loss of sovereignty of peoples and countries over food and the production process, through the denationali- sation of landownership, corporations, agribusinesses, trade and technology. There are already more than 70 countries that can no longer produce what their people need to eat (FAO, 2013).

Large tracts of homogeneous industrial plantations of eucalyptus, pine, African palm crops, etc., have been utilised for the production of pulp, wood or agro-energy, seriously affecting the environment, causing massive destruction of biodiversity and altering the groundwater table (Miller, 2010).

A Machiavellian alliance has been built in the South among the interests of large landowners, landlords and Creole capitalist farmers, and transnational corporations. This alliance is imposing the industrial mode of agriculture in the global South at a very fast pace and is concentrating landownership in astonishing ways. It is destroying and rendering family agriculture impossible and depopulating the countryside in our countries. This mode of farming uses intensive mechanisation and agrochemicals, evicting the workforce and causing the migration of large contingents of the rural people.

A new international redivision of production and labour is under way, which condemns most of the countries in the South to being mere exporters of agricultural raw materials and minerals.

Most of the governments, although chosen by electoral processes considered to be 'democratic', are in fact driven by the logic of capital and all kinds of media spin, which has resulted in them becoming subservient to those interests. This has also translated into their agricultural policies, which have been fully subordinated to the interests of transnational corporations (Delcourt, 2010). They have forsaken state control over agriculture and food and public policies to support farmers and food sovereignty and to protect the environment.

The model of capital for agriculture: Agribusiness

In short, capital and its capitalist owners, represented by large landowners, banks and domestic and transnational corporations, are implementing the so-called production model of agribusiness all over the world.

Agribusiness is characterised by the following:

1. Agricultural production is organised into monoculture (single crop) in increasingly larger areas.
2. There is intensive use of agricultural machinery, at a progressively larger scale, evicting labour from the countryside.
3. Agriculture is practised without farmers. There is intensive use of agricultural poisons, agrochemicals, which destroy the natural fertility of the soil and its microorganisms and pollute groundwater; even the atmosphere is polluted when defoliants and desiccants are used, which evaporate and then return with the rain. But above all, the food produced gets contaminated, resulting in grave consequences for the health of the population. More and more GM seeds are being used, with standardised production techniques that seek only the highest profit rate in the shortest amount of time.

This production model that seeks to produce dollars and commodities, and not food, has become dominant and, to an increasing extent, has also been using fertile land for the production of agro-fuels for 'feeding' fuel tanks of automobiles, and is engaging in industrial plantation of homogenous trees for pulp (for the packaging industry) and energy in the form of charcoal (GRAIN, 2007).

The Contradictions of Capital Control over Agriculture, Especially in the South

The description of economic power over agriculture, nature and agricultural products scares everyone. And it can lead to pessimism about the possibility of reversing this situation, such is the force that international and financial capital exerts. However, all these economic and social processes bring with them contradictions. These contradictions generate riots and anger, adverse effects that will continue to return in the medium term.

Some of these contradictions of capitalist control over agriculture and nature are highlighted here, so that one can understand them and act on them, bringing about the necessary changes.

- The production model of industrial agriculture is totally dependent on inputs, such as chemical fertilisers and oil by-products, with their natural physical limitations like shortage of global oil, potassium, lime and phosphorus. Therefore, its expansion is restricted in the medium term. And its cost-to-price ratio is above the actual value.
- Oligopolistic control by some companies has raised food prices above their value, which will lead to hunger and unrest among the population that cannot access the food due to lack of or insufficient income. That is, simply conditioning food to profit rates will bring grave social problems in the short term, since the poorest, starving and hungry population will not have enough income to become consumers of foods that have become mere commodities. FAO (Food and Agriculture Organization of the United Nations) revealed that more than one billion people go hungry every day (FAO, 2013) – we have reached this magnitude of hunger for the first time in human history, Meanwhile, the production of food grows systematically.
- International capital is controlling and privatising the ownership of natural resources, represented by land, water, forests and biodiversity. This affects national sovereignty and will provoke reactions from a wide range of social sectors that oppose it, not just peasants.
- Industrial agriculture is based on the need for an increase in the use of agrochemicals as a way to save labour and to produce by means of large-scale monoculture. This produces contaminated food, which affects the health of the population. People living in cities, who have more access to information, will certainly react. The wealthy classes are already protecting themselves, and in large supermarket chains the consumption of organically produced foods is constantly on the rise.
- Large-scale production evicts labour from the countryside and, as a consequence, there is an increase in the population living on the outskirts of large cities (Delcourt, 2010). These people have no employment alternative and income. This increase in social inequality and rural exodus worldwide reveals a contradiction in capitalist control over agriculture.
- Companies are expanding agriculture based on GM seeds. But at the same time, there is an increase in the number of negative reports about the consequences of GM crops on the destruction of biodiversity, climate and the threats to human and animal health. Nature's reactions to this

homogenisation of plant life are becoming increasingly clear. GM seeds contaminate other seeds and cannot coexist with other similar species. Moreover, new diseases emerge in plants that are resistant to the poisons used in combination with the GM seeds (GeneWatch UK and Greenpeace International, 2008).

- Monoculture industrial agriculture systematically destroys all biodiversity. The destruction of biodiversity alters rainfall patterns and climate, and contributes to global warming. This contradiction is unsustainable and the population living in cities will begin to realise and demand changes.
- The privatisation of ownership of water, whether rivers, lakes or groundwater, will increase its price and restrict its access by low-income populations, with grave social consequences. In several countries in Latin America, the three biggest corporations in this segment are Nestlé, Coca-Cola, and PepsiCo, and together they control most of the market for bottled drinking water (CETIM and GRAIN, 2012).
- The increased land acquisition by foreign corporations and the uncontrollable denationalisation bring contradictions for the political sovereignty of countries.
- The expansion and use of industrial agriculture to produce agro-fuels will further expand monoculture and the use of oil-based fertilisers. This will not solve issues of global warming and carbon emissions. The main cause of these problems is the growing use of individual transport in cities fuelled by the greed of auto companies. Therefore, fostering the agriculture of agro-fuels will not solve the problem; it will only aggravate it leading to the destruction of biodiversity.
- The project of international redivision of labour and production turns many countries of the South into mere exporters of raw materials, and undermines national development projects that could ensure employment and income distribution for their populations. This will generate income concentration, unemployment and migration for countries of the North.
- The agriculture companies, coupled with financial capital, are also advancing towards concentration and centralisation in supermarket distribution with global oligopolistic networks like Wal-Mart and Carrefour. This process will destroy thousands of small stores and local merchants, with incalculable social consequences.
- Industrial agriculture must increasingly use hormones and industrial drugs for the mass production in the shortest time of animals for slaughter, such as poultry, cattle and pigs (CETIM and GRAIN, 2012). This will have adverse consequences on the health of the consumer population.
- Large landowners are no longer in control of the production process and profit margins. They are hostages of companies that control production and trade (Stedile and Görgen, 1993). Therefore, most of the profits

remain in the hands of trade companies. To compensate for the split in the profit rate, agriculture capitalists have increased the exploitation of wage workers, imposing a seasonal work system, by which temporary employment is available for only a few months in the year (Stedile and Fernandes, 1999). In several countries, work practices analogous to slavery, or the super-exploitation of labour, have re-emerged, with wages that are insufficient to guarantee survival and workers in constant debt to the 'bosses'. There has also been a rise in the exploitation of female and child labour (UNICEF, 2009), especially during periods of harvest when there is high labour demand, stimulating the migration of temporary labourers without assuring them any social rights (Social Network for Justice and Human Rights, 2007).

- In the model of domination used by capital in agriculture, there are no jobs and income alternatives for the youth. This is a huge contradiction, since if a productive sector cannot rely on the youth, it will have no future (Caldart, 2000).
- Vast regions within countries are becoming depopulated, and it seems as if human survival is dependent on concentrating the population in large cities (Delcourt, 2010). And there, in such demographic concentration, the living conditions deteriorate even further. Agriculture is being practised without people. The best example of this contradiction is the United States, where the prison population is greater than the population living in the rural areas.

A New Peasant Programme for Agriculture

In the literature on political economy and sociology, there is much confusion about 'peasant' as a term and concept (Evenson and Pingali, 2007). The term is usually used in association with forms of production of the past, to refer to the pre-capitalist class of farmers. In the history of industrial capitalism, capital used different forms of coexistence with and exploitation of peasant farm work in its logic of accumulation. In general, it combined dialectically both the destruction of peasant forms and their reproduction.

In La Via Campesina (the International Peasants' Movement) we have accumulated debates and theories that propose a new model of agricultural organisation, based on the hegemony of rural workers who live in peasant-like conditions (see, for example, Stedile, 2005, 2006, 2007a, and Nicholson, Montagut and Rulli, 2012). But the ways of organising this new model depend on the objective conditions of the productive forces and the nature of each country, as well as the degree of social expression of this segment of workers.

We call it a new programme because it is actually a popular anti-capitalist programme – the anti-model of capital control, a new production model under workers' control – to produce according to the needs and rights of all people.

It is virtually impossible to systematise the proposals that peasant movements in each country have defended on a single principle – a universal alternative platform for an agricultural model – since each country has its natural specificities in terms of its productive forces, class and power correlation (Stedile, 2007a).

The list that follows represents a summary of what has been proposed by peasant movements within Latin America for a new organisation model of agricultural production in the countries of the region. The main proposals are:

- *To implement a programme for agricultural production and hydropower that prioritises food sovereignty and the production of healthy foods for every country.* This means that states must develop policies of incentive and support, enabling each region of a country to produce the total amount of food its people require, thus achieving food sovereignty in the entire country. To ensure food sovereignty of the people must be the main objective and the first priority of any programme of agricultural and rural development. International agricultural trade between countries should be reduced to exchanges of the surplus or complementary staple products, acknowledging people's diverse eating habits. This should be the main goal of the organisation of agricultural production in every country and in all countries of the world.
- *To prevent the concentration of private land, forests and water ownership, and to organise a broad distribution of the largest farms, establishing a maximum size limit for the ownership of nature's goods.* The essence of the agrarian reform should be a broad democratisation, for workers, peasants and the population living in rural areas, of the access to landownership and land use, as well as to water and other goods of nature.
- *To adopt systems of food production based on agricultural diversification.* Monoculture destroys natural equilibrium and imposes the use of pesticides. Practices of diversified agriculture must be developed in all areas. There must be production and work throughout the year and this must happen in a balanced way respecting biodiversity and the environment.
- *To adopt production techniques that seek to increase the productivity of labour and land, with due consideration for the environment and biodiversity, and to fight the use of agrochemicals, which contaminate food and nature.* In general, these techniques have received the designation of agroecological practices. However, in each country different terminology is used to explain similar methods of production.

- *To develop the organisation of agricultural industries on small and medium scales, in a cooperative manner under the control of industrial workers and peasants who produce the raw material.* Agro-industry is needed in the modern world in order to preserve foods and transport them to cities. But we must ensure that agro-industries are under the control of workers and peasants so that the income from the value added to products is distributed among the workers. Also, adopting a smaller scale allows for an easier dissemination to all regions and rural municipalities, generating in rural areas more employment and income opportunities for young people, who are more open to working in such agro-industrial ventures.
- *To adopt agricultural machines to reduce workers' over-exertion, but which are best suited to the environment, and must therefore be of small scale and adapted to peasant agrarian structures of small and medium-scale production.*
- *To prevent foreign companies from controlling food production, production of agricultural inputs and food in any country.* These should be controlled by social forces in that country, whether that is the government, business workers, rural workers or peasants.
- *To defend a 'zero deforestation policy', protecting nature and using appropriate natural resources that favour those who live in the area.* It is possible to produce the necessary food for the local population in all countries of the world without the further destruction of a single hectare of forest biome or native vegetation. In addition, governments must promote massive reforestation plans using native and fruit trees in the already degraded areas in every country.
- *To preserve, disseminate and multiply native and improved seeds, according to the specific climate and biomes, so that all farmers have access to them, and to prevent the spread of GM seeds.* Farmers have the right and duty to produce their own seeds, control them and to have access to technologies that can improve them genetically, adapting them to local biomes. They must also be allowed to search for greater productivity.
- *To ensure that access to water as a common good is the right of every citizen.* It cannot be treated as a commodity and must be accessible to everyone. Aquifers (underground water) and all naturally existing sources of water in our countries must be preserved. Similarly, states must develop policies to reforest riverbanks and lakes, to protect water springs, and to provide proper storage for rainwater.
- *To implement a popular energy plan for each country, based on energy sovereignty, and to ensure that the control of energy and its sources is with the people.* This means that every town, municipality and region of our countries can develop for its needs and uses the production and

distribution of energy from renewable sources, non-damaging and non-predatory, such as agro-fuel, hydropower, wind and solar energy.

- *To ensure legal recognition to all native communities, to their cultures and to their rights of possession and use of indigenous and traditional lands or territories.* In all countries there are many native communities that, according to the local culture, are called indigenous peoples, native communities or autochthonous communities. In Brazil and other countries that have suffered under slave-labour plantations, many communities remain of the descendants of African slaves who have lived in the occupied territories for decades, but are not legalised. These communities have resisted all forms of advancement from private property and capitalism. It is essential to build a new agricultural production model for democratic occupation of the territories, whereby all these communities have assurance by the state of their historical rights over goods of nature and the territories and lands they occupy.
- *To prohibit any foreign company to own land in any country of the world.* As part of the internationalisation of capitalism through transnational corporations sponsored by financial capital, there is a land purchase rush in most countries of the South by imperialist companies of the North (Miller, 2010), or sometimes even by large companies in the South operating in mining, hydroelectric plants, pulp, etc. It is essential to ban the denationalisation of land use and the ownership of land and nature's goods (such as water, biodiversity, minerals) by these foreign companies. People's sovereignty should be protected, preventing control of their territories by foreign companies in any country.
- *To promote the development of public policies for agriculture through the state by assuring the following:*

  1. priority of production of food for the domestic market;
  2. profitable prices for small farmers, guaranteeing purchase through various state or social mechanisms;
  3. a rural credit policy, particularly for investment in small- and medium-sized agricultural businesses;
  4. a state policy to control agricultural and husbandry research, prioritising research on food production and agroecological techniques that provide broad access to farmers, and democratising their findings to the entire population;
  5. sanitary regulations of agro-industrial production adequate for the conditions of peasant agriculture and small agro-industries, thus expanding the possibilities of food production;
  6. appropriate public policies for agriculture according to the regional realities of each country.

- *To ensure social security policies for the entire rural population, as well as a public and universal system of solidarity for all workers in order to have access to health, social welfare and retirement services.* In most countries, peasants and (temporary or permanent) rural workers are excluded from public health systems and social security that provide the possibility of retirement and social assistance. Therefore, it is essential to universalise these services through appropriate social security policies for the entire population. The historical gains of the working class after years of long struggles in the twentieth century should be extended to all rural areas.

- *To review and modify the current model of individual transport in force in most countries, which is highly pollutant and can lead to distortions due to the production of agro-fuel.* A national public transport programme must be developed to prioritise rail systems, subways and waterways, which require less energy and are less polluting and more accessible to the whole population. This condition will allow for the development of more rational agro-fuel policies that will prevent large tracts of land being shifted from food production to fuel production for private automobiles, as is the case with the current production of ethanol and biodiesel.

- *To promote education in the countryside for everyone; to ensure the implementation of a broad educational programme in rural areas, which is inclusive of the reality of each region and aims to raise the social awareness of peasants; and to universalise the education of young people at all levels, in particular, high school and university, and to develop a massive literacy campaign for all adults.* The programmes giving young people access to university must be combined with housing in rural areas and designed in rotation, articulating theory and practice, in order to avoid higher education becoming an encouragement for rural exodus. Instead young people must be motivated to apply their university knowledge in their rural communities.

- *To change the current international agreements of the WTO, European Union and Mercosur/Mercosul conventions and United Nations conferences, which only promote the interests of international capital and free trade to the detriment of the interests of peasants and the people from the South.* The current agreements merely reflect the needs of capital accumulation and control over the production of goods and world trade, and are conducted by governments that only represent the interests of capital. It is necessary to break these unlawful impositions and create a new landmark for international representation, which will ensure the representation and the interests of the people.

- *To adopt the production of pulp and paper in smaller-scale industries to meet the needs of the local people and avoid the extensive monoculture of large homogeneous tree plantations that upset the balance of the environment.*

- *To develop policies to improve living conditions in villages and rural communities, ensuring access to electricity, transport and housing appropriate to their micro-climates.*
- *To encourage all social relations in our societies to be based on nurturing values shaped by humanity over millennia, such as solidarity, social justice and equality.* These values are not merely statements of principle, but should guide our daily behaviour, our movements, organisations, political regimes and states. Society will only have a future if it cultivates the historical humanist and socialist values. All societies based on individualism are doomed to failure.
- *To defend and enhance the cultural habits of each village and community as a political and cultural resistance against the standardisation imposed by capital.*

Political and Organisational Challenges for Peasant Movements in Latin America

The advent of the new phase of capitalism, in which its companies and corporations have become international, has brought with it a contradiction: it has forced peasant movements, which are in general more concerned with local and national themes, to become international too. Thus, since the 1990s, initiatives and networks have multiplied among various peasant movements in the world. These networks have resulted in the constitution of the Latin American Coordinating Committee of Rural Organisations (CLOC) and other similar initiatives in Europe, Africa and Asia. From this movement, La Via Campesina was born as a network for unity and international exchanges of experiences, principles, debates, ideas and the building of joint mobilisations in order to face a common enemy in the international arena: transnational corporations, GM seeds, and international agreements, such as those of the WTO and the World Bank, which are only in the interest of capital and implemented against the peasants (Stedile, 2007b; Nicholson, Montagut and Rulli, 2012). From these exchanges, collective reflections and experiences that have been accumulated in bilateral meetings and international conferences, we can surmise the main challenges still facing the peasant movement in the international arena today, particularly in Latin America. They are common to all countries, but must be tackled at the national level by each nation's own movements. The challenges are:

- *To transform the struggle for land into a struggle for territory.* The struggle for land is no longer a mere struggle of the peasant family for a space to work, produce, survive and reproduce. It has become more than an individual need and must be addressed as a collective need of all

communities in defence of their territory. The peasants as a class must
defend their territorial spaces against the interests of capital in order to
survive. In earlier times, the fight around land was aimed at eliminating
ground rent and the exploitation that landless peasants suffered at the
hands of large landowners and landlords. Today, capital is fighting for
land in order to control seeds, water, biodiversity, minerals, rivers and the
production of agricultural commodities. Thus, the struggle for agrarian
reform should be carried forward by all categories of peasants and rural
workers, not just by the landless.

- *To build a new model of agricultural production managed by workers
  and peasants.* Historically, peasants have been used to defend only their
  immediate interests (Casanova and Herrera, 2014). Therefore, they fight
  for land, better prices and better living conditions in their communities,
  which are measured by improvements such as electricity, roads, schools
  and other public services. Today, two models for agricultural production
  are at stake. How are we going to use our land and territories? Are they
  destined to serve the capital accumulation of some companies that
  only exploit them to produce goods and to profit from nature? Or are
  we going to allocate a social function to them that benefits those living
  in rural areas and the entire society? Thus, there is a dispute between
  the two models of occupation and use of land and territories, and they
  are incompatible. We are aware that the model of control of capital over
  production and nature puts at risk the very survival of biodiversity,
  nature and humans; it is predatory and socially irresponsible, and it aims
  at quick and easy profits. It will have serious consequences for the balance
  of the environment and human health. Consequently, it is imperative to
  defeat the project of capital for agriculture.
- *To address the interests of transnational corporations and their control
  mechanisms.* Earlier, during the mercantile and industrial phase of
  capitalism, the main enemies of peasants appeared to be large landowners,
  the local oligarchies, and intermediary traders, who exploited farmers
  and prevented them from reproducing as a class. Nowadays, there is
  a new class of common enemies of peasantry in all countries: it is the
  transnational corporations, which control territories, productions,
  technologies, inputs, and prices and the world market of agricultural
  goods. These companies operate in partnership and are sponsored by
  financial capital. Therefore, the new and powerful common enemy of all
  peasants around the world has spread. Peasants need to identify it and
  act to stop its advance, as a condition not only of improving their living
  conditions, but of their survival as a class
- *To build a new technological matrix based on agroecology.* During the
  twentieth century, peasants were generally misled by the intensive

campaigns of industrial capital that the only way to increase labour productivity and cultivated areas was the intensive use of inputs produced by industry: machines, chemical composts, fertilisers and agrochemicals. Throughout the century this production matrix was developed based on chemical products and machinery from the industry at ever-increasing scales. Many peasants were deceived into adopting them. They did not realise that by embracing the technological matrix of capital, besides having to work to pay for it, they were becoming similar to capitalist farmers. However, when their production methods were compared, they were not able to match the scale of the capitalists.

This resulted in deficits, bankruptcies and loss of land by millions of peasant families worldwide (Amin, 2005). Peasants urgently need, in all countries, to develop a new standard, a new technological matrix for agricultural production that allows for an increase in productivity of labour and yield of cultivated crops in equilibrium with the environment – to produce more, but in a healthy way. This technological matrix is summarised in the techniques brought together by agroecology.

Nevertheless, in order to do that, we need an enormous effort to collect the practices and knowledge from popular wisdom that has been in existence for decades and has been passed on from generation to generation in our communities. We need to systematise these scientific findings, aggregate them, and develop agronomy courses based on agroecology. Most agronomy universities and colleges have been taken over by the interests of capital and are contemptuous of agroecology as an important branch of science. It is up to peasants and their organisations to recover and systematise this knowledge, organising university courses in agroecology in all countries in order to give a scientific basis for a new productive matrix, which benefits farmers and the society and also maintains the equilibrium of the environment (Caldart, 2006).

Hence the importance of the efforts in which La Via Campesina is engaged all over the world, particularly in Latin America, in partnership with several progressive governments as well as university professors aware of the importance of organising and multiplying in our universities agroecology courses that are accessible to the peasant youth farmers. This connects them with networks in the continent within the Latin American Institute of Agroecology (IALA). We must make an effort to have in each biome of our countries courses in agroecology that prepare agronomists and systematise a production matrix adapted to each region.

Therefore, we need new networks of knowledge and of appropriation of production techniques in order to implement them in rural development programmes. In that sense, we can underline the importance of the experience that Cuban peasants have acquired from the methodology for the dissemination of knowledge and techniques of the Campesino a Campesino (Peasant to Peasant) Movement, in which peasant leaders themselves are encouraged to create conditions to share knowledge and experience with farmers from other regions (Holt-Giménez, 2006). Furthermore, it is also necessary to develop new methods for dissemination of agroecological techniques.

- *To implement and support schools at all levels in the countryside.* The access to knowledge is as important as having land, controlling territory and producing goods. Knowledge is the only thing that truly frees people. Knowledge is culture. Accumulated knowledge is science that humanity has been amassing to understand and transform the world. Therefore, it is essential for peasant movements and people living in rural areas to have access to knowledge. Knowledge is ordered in our societies through books and schools. Peasant movements have to transform schools into ideological territories of class, to incorporate them into their programmes of struggle, to have schools at all levels for young people and adults, from elementary school (up to the eighth grade) and secondary school to higher education and university. Schools have to be situated where people live. We must avoid programmes that move our youth and children to the city, as many governments advocate. This destroys rural roots, imposes enormous sacrifices, and slowly alienates the youth from their environment and social class. We must fight for programmes and books at all educational levels that are appropriate to the needs of our people; for teachers and educators that are in tune with the interests of the people; for public and free education in rural areas and society, under the responsibility of the state; and for these things to be considered as rights assured to every individual.

- *To develop an ongoing training process for the grassroots, militants and cadres.* Peasant movements urgently need to invest all available energy, and human, economic and material resources, in creating the necessary conditions for the development of training programmes. Training means to have class awareness combined with scientific knowledge, and training programmes help us to use the scientific knowledge developed by humanity to interpret the reality we live in and to enable us to transform it. Without scientific knowledge, or study, it will be impossible for peasants to interpret reality and transform it in the correct way. Therefore, it is necessary to develop training programmes at various levels:

   1. Mass training at the social basis for all age groups and with organisations offering employment services. In general, mass training is imparted by practising being part of mobilisations, massive forms of struggle, and making use of the media. Another

possibility for mass training is the use of cultural expression, such as theatre, music and painting.

2. Training in small clusters in an organised way, that is, basic training.

3. Training of activists, aimed particularly at young people to prepare them to be active agents and disseminators of ideals, programmes and actions. Activists compose the active body of our movements.

4. Training of leaders, which requires a higher and more complex level of scientific knowledge about the current situation of the struggle between agricultural production models.

To develop these various levels of training, it is necessary to use a broad range of forms and methods according to the culture and idiosyncrasies or specificities of each region and nation.

• *To develop our own means of mass communication.* Class struggle in the current phase of finance capitalism and globalisation is increasingly involved in the use of mass communication. The ruling classes in our countries as well as internationally have complete hegemony over mass media – television, news agencies, radios, newspapers and magazines – and use them to reproduce their ideas, ideologies, projects and programmes for society. They use them to fight against the working classes, to disseminate untruths, to affect the thinking of the masses, and to manipulate the masses of workers in the countryside and the cities (Herman and Chomsky, 1988).

It is vital, therefore, for all worker and peasant movements to develop their own media. We must not debase ourselves by speaking in the dominant class's language. Although in adverse economic and technological conditions it is essential for us to have under our control the most diverse means of communication with the people – local news, community radio, television, newspapers, etc. –, we must also develop other media that generate a real dialogue with the population and use all forms of cultural expression to spread our ideas and programmes among the masses.

• *To potentialise mass social struggles.* The strength of farmers' organisations is not measured by their programmes or by the fairness of their proposals and ideas. Their strength is measured by their ability to mobilise large numbers of people around the same objectives. And to mobilise many people is to conduct mass struggle.

Our enemies are becoming more powerful. Nowadays, we do not only face the rural oligarchies and backward landlords, but the large international capital and its corporations, banks and puppet governments, when they defend their interests. It is only possible to confront these dominant interests and economic power concentrated by capital with great mass strength. Therefore, peasant movements more than ever need to develop a new methodology for grassroots organisation and work towards drawing together the greatest possible number of families and make them aware of the necessity for mass struggle.

Only mass struggle can face capital, halt its offensive on our territories and start securing better living conditions for the people. Meetings, hearings, negotiations and representations are useful, but will be inefficient if not backed up by the power of the mobilised masses. In each country, we must discover and develop the many forms of mobilisation and mass struggle, demonstrating the accumulation of power and organisation to defend the interests of the peasantry and to build a new agricultural production model that serves the interests of the society as a whole.

- *To build national alliances with all categories of rural workers, peasants and people living in rural areas.* In all countries, there is a huge variety of peasant categories and workers who live in rural areas. Diversity is the result of the much differentiated development of capitalism in each region or country, which goes on reproducing different and more complex social relations. Thus, in most of our countries, we have remedied peasants, landed but poor peasants and landless peasants. In terms of categories of rural worker, there are those with steady employment, temporary rural workers, seasonal workers, and an ever-growing segment of workers known as subproletarian or even lumpenproletariat (Stedile and Fernandes, 1999). There are huge challenges involved in discovering the common needs of these different social categories living in rural areas, and in developing alliances around programmes and common forms of struggle. A single section of peasants, no matter how determined and radical, will not be enough to face the power of the enemy. We must always remember that the biggest challenge is to accumulate social power; and social power is the number of people organised around the same goal.

In many countries, there is also a need to build alliances with other social sectors living in rural areas, which do not identify themselves as social categories of capitalism, namely, indigenous peoples, native communities, Afro-descendents, populations living on riverbanks, and fisherfolk.

- *To build alliances with city workers.* Social changes in our countries will only be possible and feasible when we can build a broad mass movement bringing together the entire working population from the countryside and the city. No social force alone will make the necessary changes for the entire society. We need to build a major national alliance among all the working classes and the oppressed and exploited peoples.

There are two classic ways to go about building this major and necessary alliance:

1. With the development of common struggles around issues that concern everyone. For example, problems of workload, education, employment, income, public services, public health, agrochemicals and environment are issues that affect the entire population. Therefore, developing forms of struggle around them may bring together broad masses.
2. With the establishment of a national programme by the working classes and the people for the country, representing a single political project.

Thus, peasant movements must be aware of this need to break free from corporatism and sectorism in agrarian issues in order to add numbers to other categories of the working classes and the people living in the cities, and to be able to build a broad movement that can have enough power to implement a new socio-economic programme of structural changes. Peasants increasingly depend on alliances with the city (Stedile, 2007a) to defend themselves against the exploitation and plundering they are subject to in the countryside. It is a huge challenge to break down barriers that separate those who live in the countryside and those in the cities in order to create common ties of goals, programmes and forms of struggle.

And finally, we have to articulate joint international mobilisations against the same enemies. Today, if the class enemies are articulated internationally through their banks, corporations and international agreements, it is necessary for peasant movements to develop their own international forms of articulation and mass struggle. The questions that are before us are: How to challenge price and market control of the crops if they are determined by five or six transnational corporations worldwide, such as Monsanto, Cargill, Bunge, ADM and Dreyfus. How to address the issue of agrochemicals if a few international companies, including Bayer, BASF, Syngenta, Monsanto and Shell Chemical have complete hegemony over technology and the market in all countries. How to develop a new dairy production model if companies like Nestlé, Parmalat and Danone influence the world's markets. How to protect our drinking water supplies if a few companies – Nestlé, Coca-Cola, Pepsi-Cola and Suez, for example – want to control it worldwide. How to fight against the privatisation of our seeds or genetic modifications that eliminate biodiversity if these practices are regulated by just a few GM seed companies around the world. How to face the advance of eucalyptus and pine monoculture if a group of pulp companies, such as Stora Enso, Botnia and International Paper, dominate the markets. To handle these questions, peasant movements must develop strategies and forms of popular struggle that are more and more internationalised.

3

AFRICA

Rebuilding African Peasantries:
Inalienability of Land Rights and Collective
Food Sovereignty in Southern Africa

Sam Moyo

Peasantries and Agrarian Transformation

Fifty years after Africa's decolonisation, following the unravelling of Apart-heid-inspired settler colonialism in Southern Africa, the bulk of the African continent's peasants persistently face a crisis of basic social reproduction, manifested in inadequate access to food and chronic malnutrition, and diminishing income from farming, pastoralism and related marginalised (but diversified) survival strategies.

Colonial and post-independence Africa failed to resolve the three classic agrarian questions, which constitute key elements of democratisation and national (integrated) development, namely: improving agricultural productivity (Mafeje, 2003), so as to improve the supplies of wage foods; providing raw materials for basic industrial and employment development (Patnaik, 2008); and promoting accumulation from below. This failure obtains whether in the semi-industrialised peripheral states (such as South Africa, with its racially discriminatory agrarian transition of accumulation from above), in the putatively 'successful' peasant-based agrarian economies (e.g. Kenya and Malawi), or in the fragile Sahelian pastoral regions. Variations in the mode of African colonisation imposed three different strategies of agrarian surplus extraction and accumulations that, while presenting different subregional specificities, have all resulted in the failure to resolve these agrarian questions.

This failed agrarian transition is the consequence of two centuries of land alienation and the super-exploitation of agrarian labour (on large farming estates and in the mines), which was historically most extensive in Southern Africa, as well as in 'non-settler' Africa by the systemic exploitation of peasantries' labour through intensive extraction of surplus from their production and through their malintegration by colonialism and post-independence rule into the unequal world capitalist trade regime. The key result until the 2000s has been the underdevelopment of Africa's agrarian production systems, through the subordination and super-exploitation of agrarian labour and consumers by monopoly capital. During the 1990s, SAPs intensified Africa's agricultural extroversion and unequal extraction of surplus value, including through a second diffuse and low-intensity wave of land concentration, and expansion of food imports and food-aid dependency. More recently there has been a pervasive effort by diverse foreign land-grabbing 'investors' to dispossess the peasantry in non-settler Africa and the periphery of settler Africa of their best lands and water resources, as well as to exploit their labour as direct workers, 'outgrowers' and 'contracted farmers'.

These agrarian accumulation strategies essentially undermine the social value of peasant production, based as it is on self-employed family labour and family lands, with the purpose of providing foods and other products primarily for auto-consumption. Indeed, poor peasants have been the most resilient in maintaining food production, even during the SAPs and various world commodity price crises, and even though their production has been inadequate (Mafeje, 2003) to sustain growing consumption needs. Peasant families mobilise family and other kinship labour, nurture biological (seed) and other local resources, and adopt new crops and technologies, especially locally adapted ones, to expand low energy-intensive agricultural production for their social reproduction, rather than for over-consumption in Western markets. The landless workers and peasantry have sought social reproduction in spite of the withdrawal by the neoliberal state of support to peasant farming and social welfare, and despite the persistence of unfavourable terms of trade. The peasantry's alleged technological 'backwardness' is driven by neoliberal policies, which disproportionately transfer the cost of inputs relative to commodity prices to them and reduce their realised incomes through the absence of state subsidies and protection.

The failure of the African agricultural reforms to prioritise the development needs of its vast peasantries, whose production systems are globally the most backward and have resulted in the highest levels of food insecurity, is ironically now presented as a justification for the land-grabbing deals sanctioned by African states and local capitalists. This further marginalises the peasantries, fuelling fresh political and resource conflicts over the new agrarian questions that land alienation imposes, and the sustainability of the ongoing polarisation of agrarian accumulation from 'above' at the behest of oligopolic financialised capital.

Primitive Accumulation by Dispossession in Africa: General Trends

Colonial capitalism: Land dispossession and peasant incorporation and accumulation

Africa of the labour reserves (Amin, 1973) or 'settler Africa' (mainly in South Africa, Rhodesia, Namibia, Kenya, Algeria, etc.) had by the 1960s witnessed the first African wave of extensive land grabbing by European settlers. Settler colonial states created 'large-scale commercial farming' (LSCF) systems based on private-property rights, assigned mainly to individual family-operated farms spatially segregated from the black African Communal Areas, including some 'enclaves' of agro-industrial estates heavily subsidised by the state. African peasants' land dispossession by the British South Africa Company and others led to widespread displacement and landlessness, which ensured the super-exploitation of cheap labour (compelled economically and otherwise), while destroying the peasant economies. Settler estates were also created in the Lusophone territories (Mozambique and Angola), and on a smaller scale in various migrant-labour 'sending' states (e.g. Malawi, Zambia and Mozambique). While these developments did not lead to the complete dispossession of peasant lands, such dispossession was so extensive as to undermine the peasantry (almost completely in South Africa), and it led to the creation of a migrant labour system across the region. This resulted not in 'enclavity' but in a functional dualism that subjugated labour and repressed peasant farming.

Accumulation from above through land dispossession and displacement of the peasantry, and through economic and extra-economic coercion of labour in former settler-colonial countries, epitomised the first wave of alienation in Southern Africa, from the eighteenth century until the middle of the twentieth century. Given a veneer of legality by the British Crown, European land settlement led to monopolistic control over national water resources and public infrastructural investments, buttressed by the dominant white settler ideology and state–society relations defined by the policies of private property rights and racially discriminatory investments favouring the LSCF, while undermining the remaining peasants through discriminatory commodity markets. This shifted the production of food from peasants towards wage-food commodities dominated by large farmers supported by state marketing boards and European merchants. This mode of accumulation and political rule of the Southern African state, including its institutions of taxation and the social security systems, was racially discriminatory, undemocratic and repressive, while placing the burden of social reproduction on labour and the peasantries in a subsidy on capital.

In non-settler Africa, two broad land alienation histories prevailed through an indirect mode of colonial rule (Amin, 1973; Mamdani, 1996). In 'Africa of the Concessions' (largely Central Africa), land alienation by European trading and mining companies led to the creation of a few significant enclaves formed around agricultural plantations, with rudimentary agro-processing facilities, as well as mining enclaves. The mode of primitive accumulation entailed raw material plunder and limited infrastructural investments. The pedigree of resistance to this enclave dispossession is well documented.

Elsewhere, in Africa of the 'economy *de traite*' (Amin, 1973), which evolved from two centuries of European mercantilism, there were widespread African resistances to Lord Lugard's attempts to alienate land (Mamdani, 1996). This led to pervasive growth of 'petty (agricultural) commodity production', among differentiated peasantries (Bernstein, 2002) or 'small cultivators' (Mafeje, 2003). Quite critically, this mode of colonisation also entailed institutionalised labour migration (albeit not backed by land alienation), including the incorporation of migrant farmers from northern territories of West Africa into the coastal and forest regions' economies. This led to the creation of diverse peasantries, including independent-lineage producers, farming labour tenancies and various forms of sharecropping arrangements (Amanor, 2008). Smaller-scale agricultural estate enclaves (palm oil) also emerged in various countries. Moreover, the pockets of semi-feudal agrarian structures persisted (e.g. Northern Nigeria and Ethiopia) or were created under colonial rule (e.g. Uganda). This colonialisation pattern brought diversity to Africa's agrarian transition in relation to land alienation, its agrarian structures and patterns of accumulation.

Post-independence developmentalism, neoliberalism and re-institutionalised primitive accumulation

In general, from the 1960s, post-independence governments halted the pace of land alienation and either initiated the nationalisations of alienated lands or created new leasehold land tenure systems on such restricted estates. This restricted foreign landownership and also slowed down the commodification of agricultural lands by restricting the freehold private-property regime that was being pushed by the colonial rulers. These governments also abolished the exploitative labour regimes by rescinding rural head and other farming taxes, and by reversing the institutionalised labour migration systems. Armed struggles in Kenya, Mozambique and Angola culminated in substantial but inadequate land redistribution.

Independent states sought to promote 'expanded reproduction' among the peasantry, using state marketing boards and inputs support programmes, although they tended to extract substantial shares of the agrarian surpluses purportedly for various national 'development' schemes. After independence, the dual objectives of agrarian reforms in the different African countries were to enable local state accumulation from agricultural surplus values, and through the deepening of the extroverted integration of African agriculture, to expand export cropping to increase forex revenues for the expansion of import substitution industrialisation (ISI) processes.

The 'modernisation' of agriculture was from the 1970s largely pursued through bimodal farming strategies, which sought to nurture middle- and larger-scale capitalist agricultural production systems at the expense of the peasantries, while promoting a degree of increased productivity among peasants and directing their produce towards state marketing boards; these were intended to develop national infrastructure and invest in industries. Even the national agrarian capitalists were, however, subordinated to the extraction of surplus value by transnational corporations (TNC agribusinesses), which were protected by the centralised state marketing regulation. Up to the 1970s, various African states attempted to establish a few new large-scale farming (cropping and ranching) estates, largely through state corporations, and a few individual African capitalist farmers, building mostly on nationalised colonial agricultural estates (Tanzania and Malawi), land redistribution (Kenya) and in some cases on lands newly alienated from land under customary tenures (Botswana, Malawi, etc.). Surplus extraction continued to be at the expense of the super-exploitation of African peasantries (Shivji, 2009) and through the cheap labour provided to large estates.

After being admonished by the World Bank (through the Berg Report, 1981) for failed agricultural experiments, agrarian policy bias (largely urban bias), the putative inefficiencies of state interventions (trade protectionism, state marketing regulations and participation through commodity boards) and inefficient state farming (Mkandawire and Saludo, 1999), the state retreated from subsidising agriculture.

The state agricultural estates were gradually dismantled and privatised. From the 1990s numerous domestic capitalist farming elites procured or 'grabbed' middle-sized farmlands, while a few foreign capitalist farmers and corporations established large farms in some African countries (e.g. South Africans in Mozambique and South Africa), putatively in pursuit of expanding (traditional and) non-traditional exports. Countries such as Mozambique, Tanzania and Zambia were now concessioning off peasant lands and reversing earlier land nationalisations, while Botswana, which after independence had redistributed some of its few white-owned LSCFs, was expanding its large-scale ranching by dispossessing pastoralists of their land and water resources. This second wave of land alienation led to land dispossession and the displacement of significant numbers of peasant families, albeit in scattered and smaller enclaves than those produced by the first colonial wave of land grabbing in settler Africa. This process was popularly resisted, albeit unsuccessfully, including through armed rebellion, given the feeble response of the burgeoning national 'civil societies' (Moyo, 2008; Moyo and Yeros, 2005).

The narrow preoccupation of Africa's agrarian reforms with mainstream voluntaristic approaches, which condemn the small 'farmers' for being 'traditional subsistence' farmers, despite their struggles to enhance their food self-sufficiency, was based on the presumed efficiency of the larger, 'commercial' farmers. The latter were considered more prone to 'modernised' farming and presumed to be more capable of leading agricultural transformation in Africa (Mafeje, 2003), despite their historical dependence on subsidised imported farm technologies (machinery, equipment, seeds, fertilisers and agrochemicals), and their focus on the export of agricultural raw materials, whose terms of trade were declining. This led to disarticulated maldevelopment and increasing dependence on food imports and aid.

Most agricultural transformation initiatives in Africa since the 1980s were, therefore, based on atomistic projects comprising incoherent neoliberal welfarist palliatives which, while failing to resolve the African food crisis and agricultural productivity, promoted accumulation from above on a very narrow social and radical geographic basis. Rather than enhancing the participation of the majority of small African producers, agrarian reforms mainly sought commodity-marketing and land-tenure reforms, which led to deeper integration into the global food system and prepared the ground for the current land grabbing.

During the 1990s, the commodification of land through the appropriation of land held under customary tenure systems and its conversion into private property expanded African land markets, but largely in newer 'enclaves'. The orthodox view was that the absence of clear tradable landed-property rights limited 'tenure security' and constituted a barrier to agricultural investment and food security. African countries pursued land-tenure reforms as part of the package of deregulating domestic markets and investment policies, and trade liberalisation. African land-tenure systems, wrongly characterised as 'communal', insecure and 'unbankable', continue to be identified as an underlying obstacle to agricultural development or investment into technologies which intensify productivity. Allegedly, the systems undermine 'individual' incentives and restrict the mobilisation of agricultural finance. Some African land reforms attempted to address this question through formalising and individuating land tenures (titling), establishing larger-scale (commercial) farmers and, more recently, through initiatives to 'decentralise' the 'governance' of land. Although problems of tenure insecurity abound at the local level, the thesis of land-tenure investment never found empirical grounding (Migot-Adholla, 1994), and these tenure reforms mostly collapsed. By 2004, many African countries had reformed National Land Policies, with homogeneous legal and administrative postures that enhanced land transactions (Manji, 2006; UNECA, 2004).

Unequal land distribution was generally conceived as a problem of former settler colonies (Mafeje, 1999), although the concentration of landholdings was growing elsewhere in Africa (Moyo, 2008). Land concentration emerged over time, through incremental state expropriations of land, 'formal', and 'informal' land markets, based on processes of internal and local social differentiation. Landownership inequities began to reflect class, gender and ethno-regional cleavages, as well as other social identities, which the states in crisis either failed to mediate or encouraged. Local agrarian and power differentiation emerged as class transformations broadened the base of local elites seeking to amass larger landholdings, creating growing land 'scarcity' and landlessness.

Access to land remains a problem for millions of poor rural and urban dwellers, whose basic consumption needs derive from agriculture (Amanor, 2008; Kanyinga, 2000; Kanyongolo, 2005). Restricted access to land by small producers is thus one of the key obstacles to expanded agricultural productivity and social reproduction. Until the mid 1990s, these processes represented neither large-scale land-alienation processes nor widespread landlessness or full proletarianisation (outside of Zimbabwe, South Africa and Namibia), but a socially significant and diffuse structure of land concentration and the marginalisation of substantial sections of the peasantry. This land-alienation process has gained political significance and fuels high-profile land conflicts.

In settler Africa, which includes large parts of countries in Southern Africa, extensive land expropriation and the systematic regulation of migrant labour, through organised recruitment and peasant taxation, initiated a proletarianisation process. This was done not only in the core settler economies (Arrighi, 1973), which eventually amounted more to semi-proletarianisation (Moyo and Yeros, 2005; Sibanda, 1988), but also in the eight neighbouring countries that constituted its regional periphery. Large-estate farming schemes and institutionalised labour migration systems, involving semi-proletarianisation, undermined the land rights and social reproduction capacities of labour, while subsidising capital's labour costs. The multiple social costs of expanding large-scale and plantation farming, besides land alienation, included depressed labour and income regimes, malnutrition and the marginalisation of the urban poor and peasants.

Independence of the former settler states from 1980 compromised social transformation and eschewed mechanisms for the equitable redistribution of wealth, incomes and landed property, since social changes was left to the markets and protected by 'the rule of law'. After independence, land redistribution in Zimbabwe, Namibia and South Africa was minimal, to the extent that foreign-owned farming estates actually expanded during the second wave of land alienation under neoliberal SAPs from the 1990s.

Continued malintegration into the unequal relations of the world capitalist system, including through unequal trade relations, thus entrenched domestic inequities and a crisis of peasant social reproduction (Amin, 1974). The recent volatility of and increases in the global food and inputs prices have only deepened the impoverished millions of African peasants through dispossession of the agrarian surplus generated by family labour, including the super-exploitation of women and children.

Crisis of capitalism, capitalist land grabbing and the resurgence of estate and contract farming

A major reaction of capital to the recent 'food price crisis' has been a new scramble for land in Africa, mainly to produce food and biofuels for export, using the large-estate production model (Moyo, 2008). At least five million hectares have been concessioned in over 20 African countries to foreign 'investors' (Cotula et al., 2009; Thompson, 2008; Tabb, 2008; von Braun and Meinzen-Dick, 2008). Large-scale land acquisitions through leasing and outright purchases by foreign capital in various African countries have escalated during the 2000s (GRAIN, 2009), with the explicit or tacit approval of governments and sections of the elite (Alden Wily, 2008). This represents a third wave of land alienation in all the African regions, creating numerous enclaves of large plantations or estate farming, frequently alongside perimetric 'buffer zones' of 'co-opted' small 'outgrowers'.

A new scramble over African lands for agriculture, mining and natural resources extraction is predicted, entailing a growing east–west–south rivalry to gain footholds on the entire continent (Moyo and Yeros, forthcoming). The land investors hail from as far afield as the United States and various European countries, China, South Korea, the Gulf States and Brazil (GRAIN, 2009; Petras, 2008). This trend not only raises concerns about the extent of land alienation and concentration, but also suggests the intensified subordination of the continent's peasantry and labour by monopoly capital during the present crisis.

Indeed, most of the former settler African countries in Southern Africa have encountered this 'third wave' of large-scale foreign land acquisitions, or 'grabbing', and 'investments' in agriculture, in a process which builds upon the region's already substantively privatised land-tenure regimes, based on racially skewed landownership and extensive social exclusion. The critical difference is that it is mainly previously alienated large-scale farmlands – owned by private and public corporations and individual white LSCFs – that are being sold or leased out to additional foreign 'investors'. The agrarian-accumulation model continues to be based on an outward-looking agricultural strategy – except in the case of Zimbabwe, which is veering towards internal markets, food sovereignty and autonomous development.

Social movements warn of a spectre of extensive dispossession and displacement of small-farm producers and pastoralists (GRAIN, 2009), although some 'civil society' technocracies consider these investments as holding developmental 'opportunities' and argue that the potential threat of dispossession can be mediated through internationally supervised guidelines on 'best practices'.

Some attribute these land acquisitions to a benign search for 'food security' among countries destabilised by the global food price crisis, which peaked around 2005, and putatively to the 'attraction of investment funds' to agriculture's profitability (von Braun and Meinzen-Dick, 2009); while others glorify the 'green motives' of such capital exports in search of presumably clean fuels. It is also claimed that these foreign investments are an opportunity to reverse the stagnation of agricultural productivity and food insecurity in Africa (Cotula et al., 2009; World Bank, 2008), and that they are necessary to reorient Africa's growth trajectory and to save the 'bottom billion' (Collier, 2007). Yet, land alienation in favour of agribusiness is primarily extroverted towards the production of new exports, such as biofuel, food grains, timber and tourism, which alongside the mining concessions is at the expense of the needs of existing social networks of poor- and middle-peasant households. These discourses eschew alternate endogenous agrarian reforms towards accumulation from below.

The current land grabbing is also justified by putative claims that there is abundant and unutilised land and natural resources, which are presumed to have no (formal) owners (von Braun and Meinzen-Dick, 2009). Such land alienation builds upon long-standing colonial-era attempts to 'reform' agricultural lands and natural-resources tenure systems by establishing 'private-property rights' and 'land markets', which are considered the sine qua non of agricultural investment. Indeed, the neoliberal land policy reforms unleashed during the 1990s (Manji, 2006) had resuscitated the commodification agenda and laid the legal and political basis for the current wave of land alienation.

This recent food 'supply problem' is thus being addressed through expanding agribusiness food production activities,[1] including area expansion in the South and the displacement of small food producers. These processes further divert even more financial and related resources away from small producers (Patnaik, 2008; Tabb, 2008). Most 'international financial and food-aid institutions' seek increased aid monies to lend to the food-crisis-ridden and riot-stricken poor countries for grain imports, as well as to finance more food aid. This would increase imports from the West, alongside cash transfers to the poor to buy food from abroad and from local surplus areas (e.g. South Africa). Rather than mobilising financial aid and truly concessional loans to support small farmers in order to increase food production in the South, this strategy would augment and refinance the dominance of agribusiness over food production and entrench the intensive capital–energy–food system. In this case, consumers in the SADC region remain captive food and inputs price 'takers', and provide malnourished, cheap labour to the region's 'enclaves'. This represents a form of malintegration into a dysfunctional global food system, based on the 'over-consumption' of fossil fuel energy and speculative behaviour, which undermines the 'universal right to food'.

Underdevelopment of Agrarian Production Forces

The structural distortion of Africa's agrarian system since independence is a socially constructed process that has been exacerbated by neoliberal policy regimes, which have undermined agricultural production structures and led to low levels of agricultural productivity. This trajectory did not arise from an intrinsic 'backwardness' of the scientific/technological and cultural property regimes of small producers, nor is it due to any unique 'physical' constraints (e.g. tropical soils and landlockedness) facing Africa. It was not the 'inappropriateness' of the available productivity-enhancing technologies or the absence of appropriate land-tenure systems that led Africa to be bypassed by some technologies. It was the reversal of agricultural and wider interventionist policies under structural adjustment, and the 'fiscal crises' they suffered that halted the growth of peasant productivity (Patnaik, 2008).

The absolute growth of agricultural production in Africa – including the SADC region – has been rising positively since the 1950s, albeit at a slow pace compared to trends in Asia. Maize output volumes peaked in 1981, and then again around 1996–97, only to experience numerous major dips during the 2001–2006 period. South Africa is a net surplus cereal producer, except during 2006 and 2007 when its harvests were estimated at around 7 million tonnes (compared to 11 million tonnes in the 2004–05 season), reflecting fluctuations of up to 36 per cent due to reduced areas planted and yields related to drought. Only half of the SADC countries have produced wheat since the 1995–96 season, South Africa being the main producer (peaking at 2.8 million tonnes in 1996–97 and dropping to about 1.5 million tonnes in 2003–04). The next-largest producer was Zimbabwe, which peaked only once at 320,000 tonnes in 1999–2000, to decline by 75 per cent to 80,000 tonnes in 2005–06.

However, the long-run per capita production of cereals and maize has been declining since the mid 1970s. Per capita cereal production on average ranged from 140 kg per person during the 1980s to 60 kg and 85 kg per person in 1992 and 1995 respectively. This was most pronounced in maize production, which declined from 180 kg per person in 1982 to 85 kg per person in the early 2000s.

The production of protein-rich and high-value foods (including meats, oils and fats, milk and pulses), which mainly targets middle- to higher-class markets (except in the case of pulses), is largely located in the more developed enclaves of the Southern African region. Most (about 50 per cent) of the socially differentiated small farmers do not own any livestock, but the majority produce a modicum of pulses and vegetables for their own consumption and to sell locally. The region is both an exporter of high-value beef and an importer of lower-value meats.

The 'backwardness' of African peasants' agricultural practices and technological stagnation tend to be atomistically identified as the primary sources of Africa's agricultural productivity decline. While the technological deficit is an empirical fact, it is the cause of it that is in dispute: is it due to endogenous factors, such as peasant behaviour, or to the wider systemic effects of malintegration into the world capitalist system? The decline in the SADC region's per capita food production was a consequence of both the limited land of small producers and various on-farm production constraints, including the exploitative input and output markets and unequal trade. Low levels of state investment to support small farmers, who face extreme weather volatility, played a critical part.

The deceleration of agricultural technological transformation, through reduced per capita utilisation of inputs (improved seed, fertiliser, etc.) have constrained land and labour productivity, particularly among small producers. Fertiliser utilisation, in terms of kilograms used per hectare of arable and permanently cropped land, is also low compared to other continents. South Africa, Zimbabwe and Malawi are the relatively higher users of fertiliser, at 49 kg/ha, 30 kg/ha and 23 kg/ha, respectively. This is followed by Tanzania at 13 kg/ha and Zambia, which uses a little less, while the rest use 5kg/ha or much less. The use of pesticides in SADC countries also varies greatly (World Bank, 2008). The level of agricultural tractorisation in the SADC region is relatively low – high in Seychelles, Swaziland and Botswana, but lower in DR Congo – compared to other continents. Ox-drawn traction and hand-and-hoe ploughing and weeding dominate farming practices.

In Malawi and Zambia, productivity grew from increased use of improved seeds and fertilisers and the expansion of large-scale farming (involving Zimbabwean immigrants); and in Angola due to post-war stabilisation and increased oil revenues. The yields in these countries are still relatively lower than in South Africa however. Subsidised financial (credit) markets for small producers' inputs are scarce. Dependence on costly imported fertilisers has instead increased, while new technologies are not adequately generated locally because of limited public and private investment and global agribusiness control. As farm margins decline, especially for small producers, incomes and investment also fall, extending the cycle of low productivity. Although weather volatility has led to frequent harvest failures, efforts to invest in irrigation to mitigate this are shallow: the proportion of irrigated cropped land ranges from 31 per cent (in Madagascar) to 2 per cent (in Tanzania).

The slow rate of transformation of the productive forces within agriculture is exemplified by the low levels of land yield, or productivity, and its rate of growth. The average level of cereal yields in the SADC region is about 30 per cent below the averages in Asia and Latin America (World Bank, 2008). Livestock productivity trends are also low. In Africa as a whole, between 1971 and 1997, the relationship between the growth of areas cropped to main staple crops and the growth rate of their yields was suboptimal in terms of net productivity growth.[2]

Low-intensity input use in African peasant agriculture is consistent with the region's broader patterns of weak economic growth and development, epitomised by food insecurity, which underlies the high incidence of poverty (Dorward et al., 2009). Preferential support to large farms and exports led to uneven development, reflecting the unequal political power and economic strength of the peasants vis-à-vis large farmers and the corporate capitalist sector, within the bimodal agrarian structures promoted by neoliberalism.

The anti-developmental stance of African neoliberal policies undermined the capacity of the small producers and the state to deepen technological transformation, while SAP policies led to income deflation, through wage repression and reduced public expenditure, particularly in rural areas (to below 5 per cent of their budgets), and the raising of food and farm input prices relative to wages (Patnaik, 2008). Indeed, the state retreated from financing credit, marketing infrastructure, subsidised inputs and support for technology generation and extension, as well as from financing of non-agricultural props for agricultural production and consumption, such as rural development and social welfare (consumption) transfers to the poor, as prescribed by the international financial institutions (IFIs). The inadequacy of public investments into rural and agricultural infrastructures, such as irrigation and rural transport facilities, bulk food storage facilities and ancillary services like electricity, placed a critical constraint on the capacity of peasants to expand the production of and access to food. This alongside trade liberalisation reduced the purchasing power of the poor and restricted multipliers, such as employment and incomes, leading to repressed local demand for peasant produce and farm inputs.

Deepening peasant commodity production under agribusiness monopoly

The persistent strategy during the current crisis of capitalism and agrarian accumulation remains to deepen the incorporation of African peasantries into the global agricultural exports chain, alongside the aforementioned land grabbing. The recent philanthropic initiatives for the Alliance for Green Revolution in Africa (AGRA) purport, for instance, to support the agricultural productivity growth of 'small farmers', through the scientific generation of improved seeds in 16 African food crops, to improve marketing through access to inputs and to access to private credit and 'agro-dealers'. This strategy is embedded into capital's technological and commodity monopolies, including the monopolistic generation of hybrid and GM seed technologies, rather than their mass generation at fair cost by and for small producers. This market-led strategy of promoting peasant productivity cannot reverse the systemic sources of agrarian de-accumulation, given the limited state capacities to regulate agrarian capital and reverse unequal agrarian trade relations, while supporting small farmers' production systems. Instead, without the state, this new 'peasant-friendly' market-based green revolution deepens the peasantries' subordination to the global agribusiness oligopolies.

Given that the absolute amounts of land suitable for cultivation and grazing is limited,[3] and that prime lands are increasingly captured by transnational and domestic elites, peasant agricultural production and pastoralism are squeezed by internal population growth and externally led displacements, which force more small producers to produce on physically marginal land areas, while large farming estates monopolise fertile land and water resources. Pressure on agricultural land has resulted in rapid soil exhaustion, which exacerbates the decline of yields, overgrazed grasslands and high rates of deforestation (UNEP, 2002). Incessant droughts and floods, which the state has been unable to mitigate, and the increased siltation of rivers, undermine the wider ecosystems and degrade biodiversity, alongside the effects of monocultural farming systems. This further undermines the livelihoods of small producers, while land shortages, low labour productivity and basic consumption fuel high-intensity conflicts over land and natural resources, albeit in varying degrees, in different parts of the continent.

It has also become evident that ecological imperialism and the effects of 'North' driven climate-change agenda are increasingly marshalled against agrarian development from below. The introduction of 'carbon trading' measures through aid, which seek to reserve more African land and biodiversity for external forces, tend to further displace peasant socio-economic processes. Indeed, climate change could limit the size of maize-growing areas in the SADC region (Wahenga, 2007), and the region's preparedness for the anticipated effects is limited. The 'adaptations' may entail the relocation of peasants to areas with the agroecological potential to produce food, the possibility of construction of new infrastructures, and technologies adapted to reducing growing seasons in some areas and their increase elsewhere in relation to water losses and gains. This indicates that the peasantry will continue to be marginalised in the future, as public investments in their agrarian livelihoods remain limited.

The malintegration of SADC's agrarian production structures and inputs will continue to be driven through South African capital and its brokerage of the expropriations of land and minerals in the region by foreign capital towards an export-oriented agenda, which entrenches imperial extraction of surpluses, uneven agrarian and rural development and inequitable industrialisation, leading to the persistence of the present regime of dependence on foreign food and various class- and identity-based conflicts. Seed and fertiliser production and supply are monopolised by a few transnational producers located mainly in South Africa, while fertiliser imports from distant markets are on the rise, despite the availability of local raw materials to produce these. Similarly, output markets and trade patterns show a greater reliance on external markets dominated by large TNCs, through the ('sub-imperial') hegemony of South African intermediaries (farmers, capital and state enterprises). Meanwhile SADC regional economic underdevelopment has led to intensified out-migration to South African enclaves, while peasant farming systems are undermined by cheap food imports from South Africa, Brazil, Europe and Australia.

Unequal Trade, Demand Compression and Agrarian Crisis under Neoliberalism

The recent global food price crisis and the agrarian crisis

Mainstream debates on Africa's alleged failed agrarian transition or its 'agricultural crisis' and 'food crisis' have tended to focus narrowly on the presumed physiocratic limitations, land tenure deficiencies and the putative technological backwardness of peasant producers as the sources of failure, rather than on the neglect of the effects of land alienation, the super-exploitation of labour and unequal trade relations, in restricting domestic agrarian accumulation and extroversion which underlies food-production deficits. The effects of unequal trade on agricultural and industrial development in Africa up to the 1970s have been well documented (Amin, 1974), while the evolving internal class relations and alliances with capital associated with unequal exchange and the mechanisms of surplus-value extraction entailed have been noted (Shivji, 2009). The longer-term historical process of the mode of extraction economies and industries, through colonial state transfers of resources from the South, and the illogical attempts to argue that 'comparative advantage' determines agrarian development, have also been well exposed (Patnaik, 2003, 2011). Little research has been undertaken to show how the adoption of neoliberal policies from the 1990s has entrenched agrarian crisis through unequal trade relations in Southern Africa.

Trade liberalisation, imports competition and speculative capital flows destroyed various productive activities (industrial and agricultural) in the SADC region, while increasing the production and import of elite consumer goods at the expense of locally produced 'traditional' goods. This led to further 'de-industrialisation' and net unemployment in the region. Meanwhile, income deflation arose from a secular shift in terms of trade against petty producers of primary commodities (especially of peasants' food and export crops), through monopoly capitals' pricing practices, and in relation to their oligarchic control of agricultural commodity markets. African farmers had in general already been exposed to 'global competition' from heavily subsidised farmers from the North (Action Aid, 2007) and exports were subjected to punitive non-tariff barriers (Ng and Yeats, 1996). The net result of structural adjustment was wage recession and income deflation, leading to the compression of domestic agricultural demand during the 1990s. This was exacerbated in the 2000s by the 'global food crisis'.

The global food system, which is a deeply integrated and 'oligopolistic' agro-industrial complex, had for long survived a real-terms decline in food prices, based on subsidised food 'overproduction' in the 'West' (Tabb, 2008), amidst repressed food consumption and production in the 'South' (Patnaik, 2003). The recent real increase in terms of oil price has triggered the shifts in the uses of food (agro-fuels) and its prices and in the uses of land. Continued trade protectionism, subsidised exports and imposed structural adjustments that are propped up by the food-aid system were key representations of production in the South.

The rate of increase in the prices of food grains, edible oil and livestock products, particularly between 2006 and 2008, was the most dramatic upward surge experienced over the last 30 years, given that in real (US$) terms food prices had been on the decline.[4] Some argued that the price increases reflected a mismatch of global supply and demand due to increased grain consumption in Asia; the reduction of 'Western' grain stocks owing to weather-induced harvest failures; the rise of farm inputs costs induced by oil price escalation (Ghosh, 2008); the diversion of grain utilisation to agro-fuel production (von Braun and Meinzen-Dick, 2009); and commodity speculation (Wahenga, 2007; Tabb, 2008). Chauvinistic analysts attempted to distribute national responsibility for inducing price increases (Patnaik, 2008) by arguing that Asian overconsumption of grain was the problem. Others argued that since prices more than doubled because of the rising cost of oil, the OPEC countries were to blame, while agro-fuel production subsidies were used to distract the USA and EU from their larger culpability for the crisis of capitalism and its effect on food supplies. Restrictions on the export of rice and wheat by countries like Thailand, Vietnam, India, Russia and Argentina were also blamed, albeit after the increase in prices.

The use of food for agro-fuel production and oil-related increases in the prices of farm inputs, however, were central to the food price escalation (Ghosh, 2008), as these accounted for 85 per cent of the increases, despite being proximate causes of the price escalation. The agro-fuels production process is influenced by the 'political pressures' and 'security' concerns of the Western energy industry, capital funds, the science and technology industry and the aid system, reflecting 'high levels of rent-seeking strategies' led by professional lobbies and think tanks (von Braun and Meinzen-Dick, 2009), as well as the so-called bureaucratic stasis and warped incentives that drive aid officials (Bird, Booth and Pratt, 2002). The underlying driver, however, was finance capital's oil and commodity speculation activities (Ghosh, 2008; Tabb, 2008), including futures' pricing of commodities (oil, food and others), irrespective of the trend in their actual physical supply and consumption. Wider systemic mechanisms drove the underproduction of food in the South and the related food price increases, given that the global food system is embedded in financial and commodity markets.

Indeed, the recent export of capital to Africa for the exploitation of agricultural land, water, minerals and other natural resources reflects the escalation of capital's speculative tendency to accumulate by dispossession, in the wake of the collapse of the housing, energy and derivative financial markets. The effects of the 'long' crises of the oligopolistic capitalist system (Ghosh, 2008; Moyo, 2008; Patnaik, 2008; Tabb, 2008) have been to undermine the African peasantry and agriculture in general, and to depress social and food consumption. This trend can only be reversed by national and regional policies that seek food sovereignty, including by protecting land rights, access to water and control over biodiversity resources, in favour of the peasantry to prevent further dispossession.

The SADC region's food crisis and South African capital's hegemony

The decline in food production since the 1990s and the recent food price 'crisis' in the SADC region also reflects the systemic extraction of surplus by the oligopolic agrarian capital during neoliberalism, through the subregional architecture of agribusiness centred in South Africa, rather than the intrinsic weaknesses of small farmers in relation to large-scale capitalist farming. Consumers of imported foods and farm inputs in the SADC region have been captive 'price takers' of food produced in the South African and global food markets, because South Africa plays a pivotal role in shaping the SADC food system through its transmission of food producer and consumer prices, defined by 'tree' agricultural markets related to South African pricing on a world 'party' basis. Food prices are unrelated to the region's own real costs of peasant-based production and the levels of incomes (i.e. effective demand). These trends constitute 'oligopolistic' price formation processes, related to the subsidies and protection provided to the dominant global food exporters, alongside the control of food supply by South African capital to the SADC region suffering from food deficit.

The region's failure to produce its basic food requirements, due to the compression of demand arising from income-deflationary SAPs (Patnaik, 2008), means that the unequal regional food trade regime and food import dependence shapes the SADC region's agrarian system, including underinvestment in domestic food production. Since 1985, the extraversion of agriculture in the SADC region has been reinforced by the increased export of raw agricultural materials, despite their declining terms of trade and food imports. For instance, in 2006–07 and 2008–09 (non-drought seasons) the total estimated commercial cereal imports into the SADC region amounted to 1.22 million tonnes, against expected food-aid deliveries of 0.22 million tonnes and exports of 0.2 million tonnes, leaving a cereal deficit of 2.63 million tonnes. This import cost surpassed the US$1 billion mark during 2001 and 2003 (a drought period) and has since been rising by nominal US$ terms.

Food import dependence varies among countries. Botswana imports 77 per cent of its cereal consumption requirements. Angola, Mozambique and DRC have for long been high 'food-production-deficit' countries, importing over 50 per cent of their cereal consumption requirements (Glantz, Betsil and Crandall, 2007). Tanzania, Swaziland, Malawi and Zambia are 'relatively minor food-importing-and-occasional-exporting' countries, importing between 13 and 50 per cent of their cereal requirements, depending on frequent climate-induced 'harvest failures'. Since 2003, Zimbabwe has imported between 30 and 60 per cent of its cereal consumption requirements. The relative cost of food imports has been growing, placing greater pressure on the limited foreign currency resources of most SADC countries and diverting resources from other social and economic investments, including improving agricultural productivity.

Food-aid deliveries to the SADC region increased sharply from 2001 until 2007, when they returned to the 1998 levels. Between 2001 and 2003, US$1 billion had been provided (i.e. an average of $250 million per year). The proportion of the population requiring food aid (during 2001 and 2003) varied widely: 48 per cent in Zimbabwe and Zambia, 32 per cent in Malawi and Lesotho and 29 per cent in Mozambique. Food aid and import dependency mean that food prices within the SADC region are influenced by the vagaries of global markets as well as by intra-SADC trade.

Yet, between 1995 and 2006, the share of agricultural exports in the SADC region's total exports averaged 23 per cent (UNCTAD, 2008), while agricultural imports averaged 31 per cent. Excluding data on South Africa and the extreme drought years, the share of agricultural and food imports and exports rises much more in most of the countries. In terms of the agricultural trade balance (e.g. during 2004–05, a non-drought year), seven out of the fourteen SADC countries imported much more (in US dollars) than they exported. However, nine countries imported more food than they exported. This indicates that a significant share of national resources has been diverted to agricultural exports, while large amounts are spent on food imports. Beverages (coffee and tea) and spices dominate the exports, followed by sugar, vegetables, fruits and cereals. Imports are predominantly cereals, dairy products and meat preparations, indicating that although these main exports bring in US dollars, the production of high-value foods has unfortunately been delegated to the rest of the world. A few countries lead the exports, while most are heavily dependent on food imports.

While overall trade between SADC countries is low, food trade is dominated by South Africa. Five Southern African Customs Union (SACU) countries import over 70 per cent of their food requirements from South Africa, while the other countries intermittently import large amounts of grain (and larger amounts of dairy products and other minor foods) from South Africa and the rest of the world. South Africa, even under apartheid, has always been the dominant grain exporter. Recently, Malawi and Zambia have been exporting significant amounts of maize. Grain supply in these three countries has tended to influence regional food price formation, by repressing maize prices when they have enough to export to the region, and vice versa. The time and costs of transporting traded goods (including food) within or among SADC countries, due to limited transport infrastructure investments, has been a key constraint to both balanced regional integration and collective approaches to addressing the region's agrarian (food) production and food access deficits, particularly during extreme droughts.

Food price formation and trade in the SADC region involve the transmission of global prices in food and farm inputs through South African pricing processes, since the uneven and erratic structure of regional food production and marketing within the region enables South African agribusinesses to dominate regional food markets. Indeed, recent increases in food prices in the SADC region were inordinately influenced by the global food and energy crisis, in spite of the fact that its food consumers and its predominantly small producers are among the lowest users of farm inputs and related energy resources.

SADC food prices have risen rapidly, albeit not in the same way as in the global trade markets. South African bread prices increased substantially, although wheat (producer) prices moved at a much slower pace. This suggests that South African food processors (agribusinesses) were taking the lion's share of the price increases. Since South Africa is the dominant supplier of food and farm inputs in the SADC region, its system of price formation of agricultural commodity and inputs is likely to influence prices in the SADC region, largely because these food and farm inputs tend to be parity priced.

Thus, South Africa is both a transmitter of global prices and a sub-hegemonic pacesetter of food prices in the SADC region. Its food prices rise or fall in some sympathy with global trends as well as with the volatile regional food production balances occasioned by frequent droughts.

South African food-producer price hikes since 2001 initially acted independently of global food prices by increasing sharply during the extreme drought-induced food-grain deficit in SADC (between 2001 and 2003), and due to speculation on the rand in 2002 (Roberts, 2008). Only later, from the 2004 and 2005 seasons, did the prices follow the dramatic global food price hikes. This was possible because the entire share of the SADC region in global grain output is low, while South Africa's share in the SADC market is dominant.

The latest South African maize price increases cannot be attributed only to a domestic supply and demand mismatch, but also to the effects of global prices and the frequent SADC region's maize production deficits, including some degree of food imports into South Africa.

South Africa is also bedevilled by global financial speculation on its stock exchange and financial markets. Moreover, there have recently been allegations of collusive price-fixing by the 'oligopolistic agro-industrial corporations' in South Africa, including on basic foods such as maize and basic inputs such as fertiliser (Roberts, 2008). These agrarian price trends are undergirded by South Africa's neoliberal economic policies, whose wider ramifications have been to repress the own-food-production capacities of small producers. Indeed, some NGO food security technocracies (Wahenga, 2007) uncritically accept the prevailing logic of the world's grain trade being dominated by the United States and the European Union, and the transmission of this food regime through South Africa's subregional hegemony. They argue that national food security can be more 'efficiently' achieved through freer food trade (largely from South Africa), rather than through national state interventions in agriculture, despite their conceding the negative effects of the recent diversion of food-grain exports to agro-fuel production. Yet, while the SADC region's share of global food production is below 2 per cent, it has increasingly become a net food importer. Moreover, the recent SADC Free Trade Area, which only allows for the protection of some 'sensitive' agricultural products (15 per cent), is not accompanied by 'developmental' support for small-farm producers and technology generation. A regional food sovereignty strategy, which avoids the displacement of local food production and control of the agrarian distribution system, is effectively constrained by this open regionalism.

Under-Consumption of Food and Increased Poverty

Neoliberal debate on the causes of food insecurity in the SADC region has mostly focused on 'internal' factors, including the inadequate implementation of SAPs and the ineffectiveness of state interventions in agriculture due to the neo-patrimonial political system (Bird, Booth and Pratt, 2002). During the 1980s, the reigning 'national food self-sufficiency' policies focused on raising domestic capacities to produce virtually all national food requirements and supply them at stable prices, since food imports were perceived as both economic and national security risks. National food reserve stocks were kept to stabilise prices and supplies, especially to combat droughts. It was assumed that adequate national food production would translate into availability and access at the household levels, including among the poor. Food self-sufficiency, however, was hardly achieved in most SADC countries at that time (except, at times, in

South Africa and Zimbabwe), and even when there were grain surpluses, these could be 'sitting' among the malnourished, as continues to be the case today, even in countries where food 'surpluses' are exported (e.g. South Africa, Zambia and Malawi).

From the 1990s, when SAPs were adopted in practically all the SADC countries and state interventions were rolled back, the neoliberal 'food security' policy framework entailed two competing aspects: national and household food security. What distinguished the 'food security' approach from the self-sufficiency approach was that the former claimed to be 'accommodative' of wider processes of national and household food supply and access processes (Kalibwani, 2005). Countries were extolled to produce their own food only if they could do so efficiently and they were not allowed to spend on storing food. Otherwise, they were encouraged to import food as and when needed, given that this was considered more effective for a number of countries that were deemed to have only a 'comparative' change in producing traditional and new exports. For households the proposed focus was to ensure that the rural and urban poor established diverse means of securing incomes or cash ('livelihoods') to procure food, while only encouraging the capable farmers to produce their own food and sell surpluses to 'net food buyers'. Imports were considered less costly to the fisc and price-competitive, although they bloated government indebtedness. Keeping grain reserves at accumulating costs was considered 'irrational', and monies were to be kept aside to procure the required food, leading many countries to drain their public grain reserves.

In the event, the failure of neoliberal agricultural policies and the wider global structural impediments to achieving adequate food production in the SADC region led to escalating food insecurity instead. The availability of adequate food at the national level was partially achieved in some countries, except during severe droughts, while household 'access' to food was left to the market, and a few 'vulnerable' social groups were provided 'targeted' food aid. Export-oriented agricultural policies in increasingly liberalised economies and the removal of food production subsidies put paid to 'household food security'. Expectedly, large-scale and 'better-off' small farmers dominated the production and sale of domestic food initially and later shifted to agricultural exports. National food imports increased, while the poor hardly improved their access to food, given the deflation of incomes and loss of jobs. Household access to the available food varied depending on class-based income inequalities (Mkandawire and Matlosa, 1993).

During the drought years, 'just in time' food imports were encouraged, from both neighbouring South Africa and the rest of the world. Only recently have Malawi and Zambia exported maize, backing the Bretton Woods advice by subsidising peasants.

Consequently, SADC countries face chronic food insecurities, especially among the poor, and food production remains inadequate. Frequent cereal deficits from domestic regional production in the SADC region are common, while food price stability has been volatile. The annual volume of cereals (maize, small grains, wheat and rice) required by the 250 million people of the SADC region in 2008 was estimated at just under 30.5 million tonnes. The average level of cereal consumption per capita in the SADC region ranged from a peak of 127 kg per person in 1981 to 112 kg per person in 1999, reflecting under-consumption in terms of minimum calorific requirements per person. Annual per capita consumption has declined by an average of about 15 kg per person, even though annual population growth rates declined from an average of about 3 per cent between 1980 and 1990 to an average of 2 per cent thereafter. However, the steepest rate of decline in per capita consumption was more closely associated with the 1991–92 drought year, followed by persistently low per capita consumption for 12 years. Some projections of calorific consumption in sub-Saharan Africa (Rosegrant, 2008), which assume expanded global production of agro-fuels, suggest that intake could decrease by 8 per cent due to cuts in household food expenditures.

The consumption and production of high-value foods (meat, milk products and pulses) is relatively low. But per capita consumption of higher-cost protein-rich foods varies remarkably, with countries such as Malawi, DRC and Mozambique being at the extremely low end of the scale compared to South Africa. Intra-country class-based inequalities in access to high-protein foods are even more pronounced than access to staple foods. Chronic vulnerability to food insecurity is common, particularly among peasant populations dependent on rain-fed agriculture.

This underconsumption has resulted in a complex food and social crisis, wherein the relative unavailability and high cost of food has affected millions of people for decades. This was intensified by the 2001–2003 droughts and the global rise in food prices since then. The debilitating health and social effects of reduced consumption (calorific intakes) or changes in consumption behaviour (e.g. switching the types of food consumed and reducing the number of meals) have been long recognised. The absolute numbers of malnourished people between 1979 and 2003 in the SADC region have ranged from 18 million to 38 million at various times. Family assets have been eroded, resulting in weak resilience and failing livelihoods. Morbidity and mortality have also risen because of increased vulnerability to water-borne diseases (such as malaria, cholera and diarrhoea).

Apparently, these vulnerabilities persist because state 'interventions are poorly targeted and not addressing the main constraints or shocks of communities' and programmes are poorly coordinated (health, education, HIV and AIDS, water and sanitation), while power-related trade imbalances (against the poor) and inappropriate policies ('which discourage trade and free markets') are a problem (Fews Net, various years). The systemic contradictions of the global food regime are not considered as being the problem.

The SADC region's agrarian crisis primarily concerns inadequate food consumption by its urban and rural working peoples due to insufficient and extroverted agricultural production and unequal agricultural trade, which emanate from malintegration into speculative global food and farm inputs capital markets. This aberration is conveyed through the sub-hegemonic dominance of South African capital – which relies on malnourished cheap labour from the SADC region – based on the concentration of wealth and land (including natural resources and minerals) among racial and class minorities. International agribusiness dominates the SADC region's agricultural (inputs and outputs) and food markets through its subordinate branches of capital stationed in South Africa and the large farmers there.

Had the SADC region's food policies, especially fiscal support to agriculture and trade protection, been tailored towards the food production and consumption needs of small farmers in order to achieve collective food sovereignty and equitable regional development, the food consumption crisis would have been averted. New concepts and visions of food sovereignty, in the context of an increasingly hostile global economy, have been proposed by social movements seeking to transcend neoliberal concepts of market-led food security, although dominant global civil society alliances stick to liberal reformist notions of improving marginal rural livelihoods. In the SADC region, only a few social movements espouse the food sovereignty concept, while only the Zimbabwean and Malawian states have dramatically confronted the neoliberal agrarian framework, albeit within the wider hegemonic constraints imposed by neoliberal policies.

Agrarian Resistances Subordinated to Neoliberalism

It took protracted armed struggles to repossess land in North Africa, Kenya and the former Lusophone states, while the nationalisation of some of the dispossessed land followed independence in the former protectorates such as Tanzania and Zambia. While the nationalist project from the 1960s to the 1970s halted land alienation and the super-exploitation of the peasantries to some extent, SAPs reintroduced this trajectory as well as primitive accumulation. The reversal of foreign and minority settler domination of land in Southern Africa only began in Zimbabwe from 2000 onwards, given that market-based land reforms were a failure in Southern Africa. Instead, the liberalised agricultural policies and land tenure, including constitutional reforms initiated in Africa from the 1990s, created the conditions for the second wave of land alienation and intensified marginalisation of the peasantry incorporated by capital and prepared the ground for the recent land grabs.

The dominant responses to the recent food crises have tended to reinforce the incorporation of the peasantry into volatile global markets, extending land alienation and increasing import dependence. The expansion of South Africa-based capital into the SADC regions' food system has deepened and now spans the supermarketisation of food distribution retail monopolies, involving European capital penetration as well as the increased prices of foods and farm inputs, and a new brokerage role played by white farmers from South Africa and Zimbabwe in negotiating and managing land concessions for the production of food, sugar and agro-fuels for export by large agribusiness from the west, east and south, under oligopoly capitalist structures.

Radical responses to land alienation, the food crisis and the demise of the peasantry in Africa that are not donor-driven are few, while social-movement activism has generally been ineffective. Popular responses, particularly regarding resistance to the inequitable grabbing of land, including popular land occupations and other forms of struggle for access to resources, while mostly isolated and localised, have gained patchy momentum, given their repression by African states.

The Fast Track Land Reform Programme (FTLRP) implemented in Zimbabwe from 2000 onwards, which has led to extensive redistribution of Zimbabwe's agricultural land and the socialisation of property rights, is one instance of radicalised agrarian reform, although one within the structural and institutional constraints imposed by neoliberalism. It has expropriated large farmlands owned by over 3,000 white farmers and 20 large foreign-owned estates, and allocated the land free of charge mainly to about 150,000 poor, non-landed beneficiary families from within the peasantry and urban working peoples. Simultaneously, it also provided land to over 20,000 black 'middle-class' and 'elite' beneficiaries, while retaining some of the core lands of the agro-industrial sugar estates and wildlife conservancies. Meanwhile, the state expanded its estate farmlands from 18 to 24 and resurrected farming by the state corporation. About 20 per cent of such state farms are now joint agro-industrial ventures with foreign capital from the east combined with domestic state and private capital. Over 95 per cent of Zimbabwe's agricultural land is now state-owned and is mostly provided through land-user grants to peasants and leases to now middle-scale 'commercial' farms, while a few farms remain under freehold land rights. Most beneficiaries perceive their land tenure to be secure, with only 5 per cent having experienced evictions. Many of them are investing in the land, although some of the new middle farmers and finance capital call for private property rights in order to attract 'investment'.

Undoubtedly, fewer than expected former farm workers gained land, although in general rural labour has been relatively freed from the monopoly of the few large farm employers. In addition, the retention of the retrogressive practice of 'compound farm labour tenancy' now faces resistance from agricultural workers. Land reform has integrated the previously divided territorial authority and spatial economic barriers that desegregated peasant land from the former LSCF areas, leading to greater flow of peoples, goods and services between them. The extension of hereditary local authority into the redistributed land areas has the potentially retrogressive implication of reinforcing patriarchal relations that undermine women's land and labour rights. A key regressive feature of the disproportionate representation of middle-class and elite beneficiaries is that some of them, including those with multiple plots, argue for even larger land allocations and call for freehold property rights, while a few sublet their land to former large farmers. The consequence is a new inter-class inequality in the control of public resources and influence over agrarian policies.

In a comparative context, however, the redistribution has significantly altered property relations in terms of the relative distribution of land and the socialisation of property rights. This has created the prospect for progressive agrarian change if socially just and developmental agrarian policies, such as food sovereignty, are affected. Already, agrarian change entails the broadening of the food production base and increasing productivity among small- and medium-scale farmers. However, productivity remains low largely due to the shortages of fertilisers, irrigation facilities and draft power. Such shortages arise from reduced supply capacities of domestic agro-industrial inputs and forex constraints on imports, partly due to Zimbabwe's international isolation. The inputs shortages and new inequalities in access to agricultural inputs, public subsidies and the limited available finance have predominantly affected the peasantry. While production of food grains remains underfinanced, the recent return of agrarian merchant capital to subcontract tobacco, sugar and cotton production has reintroduced a degree of obsession with export-oriented farming.

New alliances of multi-racial domestic and foreign capital now dominate the restructured agrarian inputs and outputs markets, increasingly managed through exploitative subcontractual relations, while exposing the new farmers to unfair international terms of trade. The prices realised by the mostly small producers of maize, cotton and some oilseeds are below world prices. Current state and donor inputs support to smaller producers is minimal and provides little agricultural machinery and infrastructural investment, largely because it does not support the recovery of domestic agro-inputs industries. Private-contract farming and commodity merchants dominate agrarian markets because of the reduced fiscal capacity of the state in a 'dollarised' economic policy framework and the so-called 'illiquidity' of the financial sector, ostensibly due to the 'absence of investor confidence'. China has expanded the financing basis of the agrarian reform to fill the financing gap left by runaway European capital, but financial allocations to farming and agro-industry remain inadequate.

The Zimbabwean experience suggests that even under neoliberalism, the potential for extensive land reform in support of the peasantry exists, especially where land concentration grievances related to minority racial and foreign dominance are challenged by a radicalised nationalist coalition that involves peasant movements. However, the cross-class nationalist coalition still operates within the structure of neoliberal policies, and this soon introduces agrarian distributional biases, including those purveyed through class, ethnic and gender cleavages, reflecting a hierarchical class and patriarchal political power structure. Moreover, since capital was not totally ousted by Zimbabwe's land reform, and autonomous sources of agrarian financing are limited, internal class contradictions have enabled (the politically unaccountable) international capital to reconstitute unequal agrarian relations, using liberal domestic markets tied to the unequal global trade regime.

The Malawi case, on the other hand, represents a radical nationalist state attempt to bring about agrarian reforms in the face of the recent food crisis. It has entailed a protracted effort by the state to subsidise peasant production inputs since the 2003 drought. This has led to a substantial increase in maize productivity and the realisation of national food self-sufficiency as well as some regional maize exports, despite the fact that sections of the poor continue to face inadequate access to food and malnutrition. The Malawi experience moreover implies providing subsidies to commercial fertiliser imports dominated by oligopolic agribusiness, which has indeed deepened the incorporation of peasants into agribusiness monopolies that control agricultural inputs. The 'success' has also allowed for the continued growth of export-oriented farming among the middle-sized farmers and foreign-owned estates.

The bimodal agrarian strategy followed in Malawi, however, suggests that the peasantry can be revived based on state interventions against the will of the international financial institutions' conditionalities, when the executive and a parliamentary coalition are in favour of the peasantry's social reproduction and effectively challenge key elements of donor aid under a neoliberal regime. Nonetheless, in both the Zimbabwe and Malawi cases the retention of the wider neoliberal policy framework limits the prospects of food sovereignty, let alone the advancement of a more articulated and sustainable development model independent of monopoly capital.

The Alternatives: Collective Food Sovereignty and Inalienable Land Rights

Extensive malnourishment and food-related poverty in Southern Africa point to the failure of the entire region to resolve its fundamental agrarian question of enhancing the social reproduction of its majority peasantries. The basic agrarian production forces are underdeveloped and per capita food production has been declining, except in a few food-secure enclaves, mainly in South Africa's agro-industrial, mining and commercial farm nodes. This trajectory of disarticulated development, unequal trade relations and uneven regional development reflects the political (and policy) preoccupation with narrow middle- and upper-class consumer and export markets, at the expense of the majority poor, under the direction of monopoly agribusiness and finance capital. The recent global food price crisis merely exposed the historical deficiencies of the extractive agricultural production and distribution food system based on oligopolic financialised capital in the SADC region, as elsewhere in Africa. This process is integral to the exploitative logic of the unequal 'global' food system and the crisis of capitalism. Recent attempts to 'bail out' this inequitable global food production system, through new land grabbing and new aid conditionalities that seek to further subordinate the peasantry, can only continue the depression of food production in the South.

It cannot be expected that capitalism, specifically the interests of agribusiness and financial capital, will spontaneously promote increased African food productivity in order to enhance food security and livelihoods, by supporting the technological requirements of small producers – unless it is compelled to do so by state intervention and popular pressure. Instead, a 'Western' agricultural technological and market dependence, which submerges local knowledge and technologies through the unequal extraction of surpluses and deflationary policies, continues to be deep-rooted. Now foreign capital and domestic elites pursue a 'final push' to universalise the commodification of land and its alienation by expanding contract-farming relations with the peasants, towards reinforcing accumulation by the dispossession and displacement of peasantries, at the expense of food sovereignty and social reproduction.

The alternative we propose supports civilisation and prioritises food sovereignty and the sustainable use of resources by autonomous small producers. It includes a democracy that is inclusive and substantive, and is based on social progress. Alternative developmental approaches to agrarian transformation will require different policy choices, regarding the agricultural commodities to be produced for social gains, the (re)distribution of the means of food production (particularly of land, inputs seeds and water) and increased social investments required to sustain systemic rural development. A focus on enhancing the human resources of the peasantries is the key to restructuring the food system, through endogenous research, enhanced consumer trade protection and farmer's movements, as well as influencing agrarian policy-making and programme implementation.

The SADC states ought to pursue more holistic agrarian reforms, which reverse the decline of domestic food production and food insecurity, including exposure to external shocks and increased dependency. Such an alternative cannot be merely national in focus. It has to counter the current market-based functional approach to regional integration followed by the SADC region and instead build a regional industrial (and agrarian) policy framework that systematically reverses the opening up of the region (through trade and monetary 'harmonisation') to greater (mal)integration into the global economy. The autonomous generation of sustainable agricultural technologies and increased domestic supply of the inputs focused on domestic food and local industries are essential in order to reduce dependence on volatile external agricultural commodity and financial markets. This requires a reorientation of the SADC region's agricultural policy towards collective strategies for food sovereignty, based upon collective agricultural development initiatives.

Addressing the agrarian question in the SADC region could benefit from thoroughgoing land redistribution to small producers and the regional integration of the economies, using inward-looking strategies that build on a variety of complementarities and solidarity, including the promotion of a regional agricultural inputs and outputs markets and equitable industrialisation. The creation of state-backed sustainable food production systems and reserves to combat productivity shortfalls and drought vulnerability is critical. Stimulating regional production of key imports (grains, beef, milk products, etc.), in a process that reduces the export of raw value and increases regional employment, labour productivity and incomes, and thus expands the aggregate regional markets (possibly restraining migration), is a prerequisite for food sovereignty.

Food sovereignty requires policies that defend the inalienable land rights of small producers and are built upon a socially and economically progressive development of the peasant, with substantive democracy and social progress.

Notes

1. Tabb (2008) outlines how over 440 million hectares of allegedly underutilised land in Brazil (100 million hectares), Venezuela, Guyana and Peru (80 million hectares), the former USSR (40 million hectares) and in Africa (120 million hectares) are being eyed by offshore agribusiness ventures.
2. For instance, the area under maize grew at 1 per cent per year, while the yield in tonnage per hectare grew at 1.9 per cent. The cropped areas of sorghum and millet grew at 0.4 and 0.6 per cent per year, respectively, while their yields grew at 2.0 and 1.5 per cent per year, respectively.
3. Zambia and Mozambique have extensive areas of potentially arable land that is underutilised, while countries such as Malawi and Mauritius have extreme degrees of land shortage, with low per capita levels of arable land (UNECA, 2004).
4. Traded food prices increased by 130 per cent from January 2002 to mid 2008, and by 50 per cent from January 2007 to June 2008. Grains showed the earliest and highest price increase from 2005, although the global grain crop harvest of 2004–05 was 10 per cent larger than in the previous three years and about 9 per cent higher than the 2005–06 harvest. The prices of fats and oils increased in mid 2006, although the 2004–05 and 2005–06 seasons had recorded high oilseed harvests (13 per cent increase).

4

ASIA (I)

*Rethinking 'Rural China', Unthinking Modernisation:
Rural Regeneration and Post-Developmental
Historical Agency*

Erebus Wong and Jade Tsui Sit[1]

Modernisation and its Other

Like most of the once downtrodden colonised nations, China's key historical project of the last 150 years has been to enforce modernisation. The aim and mechanism of modernisation has generally been simplified as industrialisation, a process China has pursued since the mid nineteenth century.

Wen Tiejun portrays China's development in the last 150 years as 'the four phases of industrialisation of a peasant state' with the ultimate aim of becoming a powerful modern state to counter European and Japanese imperialism, and later the US embargo during the Cold War (Wen, 2001). The first attempt was the Yang Wu[2] Movement initiated by the Qing dynasty from 1850 to 1895; the second was the industrialisation policy pursued by the Republican government from the 1920s to the 1940s; the third was the 'state primitive accumulation of capital' practised by the Communist Party regime from the 1950s to the 1970s; and the fourth was the reform and open-door policy initially promoted by Deng Xiaoping in the late 1970s.

There had been intellectual consensus on modernisation calling out for radical social reform in China in the twentieth century. Since the 1920s, all major intellectual thought had been in agreement that China needs a thorough social overhaul. The only difference was whether the model should be American capitalism or Russian socialism. Among these radical ideas and social programmes, the rural reconstruction movement during the 1920s and 1930s, represented by Liang Shuming and James Yen, was a social initiative that has been much neglected. It is of particular relevance to reconsider this intellectual heritage in post-development China. We turn to this later in this chapter.

The marginalisation of the rural reconstruction movement was not without reason. Rural China had been stigmatised as being backward and low in productivity. According to diagnosis by the intellectuals, this was the root of China's submission in the capitalist world order. In a word, rural China needed to be abnegated in order to modernise China. Rural China, along with the peasantry, had become the Other of the modernisation project.

Nevertheless, not unlike the stigmatisation of the colonised by the colonialists, the state of being rendered as Other usually implied brutal exploitation. Such was the fate of rural China. Unlike the advanced Western countries, which had colonies to exploit and then a periphery to which to transfer its cost of development, China could only rely on internal exploitation in order to accomplish industrialisation. When it was no longer profitable to exact surplus value from the rural sector, the latter served as a buffer to absorb social risks in urban sectors caused by pro-capital reforms. Such has been the essence of China's developmental trajectory in the last 60 years. To gain a better understanding of the peasantry's contemporary situation, it is advisable to look into the detailed mechanism beyond the clichéd dichotomy of 'collectivisation' and 'liberalisation' as often represented by the two figures of Mao and Deng.

The Trajectory of China's Modernisation in the Last Six Decades

After 1949, the drive for modernisation was imperative. The desire to erase the shameful memory of being a defeated semi-colony and the anxiety of lagging behind as a backward peasant country underlay the drive for modernisation. Though established as a socialist state in 1949, socialism was not an exclusive imperative for the new regime. Even before the final victory, the new government had initially opted to orient China's development toward a 'national capitalism' under the leadership and tutelage of the state. At one point, even the possibility of introducing investment from capitalist states was not totally excluded. However, the Korean War and the Cold War had forged the fate of China's subsequent trajectory. Under the bearing of geopolitical complication, the new regime finally opted for industrialisation according to the Soviet model. However, a weak country's affiliation with a powerful ally did not usually come without a cost. One of the institutional costs of Soviet style industrialisation in China was the establishment of an asymmetric dual system exploiting rural China.

Dual system

Andre Gunder Frank (1969) challenged the 'dual society' argument, which depicted Latin America as structured by a dualism of a stagnant, backward traditional rural sector and a thriving capitalist sector. Given this, the goal of development was to modernise or assimilate the former into the latter. However, Frank pointed out that what had been happening was actually an internal colonialism in which urban sectors extracted surplus from rural areas. Latin American societies were defined by a dynamic between the two sectors that mirrored the 'centre–periphery' relationship of the developed and underdeveloped regions at the global level. In fact, the correspondence was not accidental. It originated from the same historical process known as capitalism but manifested at different correlated levels.

We can discover a similar dynamic in China's industrialisation after the 1950s, which has accounted for China's trajectory in the last 60 years (Wen, 2009). First, in order to obtain technology and industry transfer from the Soviet Union, China submitted to its geopolitical orbit. Apart from paying a heavy cost in terms of human life in the Korean War, the institutional cost was equally significant. Russian aid translated into the burden of foreign debt. Armed with a powerful industrial capacity, the Soviet Union's impetus to export its products and capital along with its political, ideological and military influence soon clashed with some socialist nations' development agendas.

China's institutions that had been transplanted from the USSR, including industrial administration, bureaucracy and the tertiary education system, remained intact and became a form of path dependency despite later delinking. In order to sustain modernisation while maintaining a high-cost 'superstructure' (institutions in general), China had to have recourse to a strategy common among developing countries. Unlike early industrialised countries, which could extract resources and surpluses from colonies or externalise institutional cost by transferring it to the less powerful periphery, the new industrialising countries had to pursue a sort of 'internal colonialism' or self-exploitation by extracting resources or surpluses from less-privileged domestic sectors, especially the rural sector. Rural collectivisation (the People's Commune) was less an ideological manoeuvre than an institutional strategy to systemically extract rural surplus at a lower transaction cost.

The state thus controlled all surplus values produced by both rural and urban labour. It was a state monopoly system for production, purchasing and marketing. The central government thereby allocated resources to expand heavy-industry-based production.

As Wen Tiejun and his colleagues summarise, before 1978 China adopted four kinds of industrialisation strategy: (1) it extracted surplus value from the agricultural sector through low purchasing price of agricultural products and high pricing of industrial products; (2) it forced the modernisation of agriculture (mechanisation and using agrochemicals) to absorb domestic industrial products through rural collectivisation; (3) it mobilised intensive and massive labour input to substitute for capital factor under condition of extreme capital scarcity; and (4) when faced with economic crises, the state tried to ride them out by transferring the redundant labour force to the rural sector through ideological mobilisation (Wen et al., 2012).

The dual structure in China's society was thus institutionalised (e.g. through the notorious urban household registration system and its discriminatory welfare system that was unfavourable to the rural population).

The exploitation of the rural was rationalised in terms of the vision of building a modern China, strong enough to counter western hegemony. Hence, it is not surprising to see that the rural sector has been appropriated for the realisation of industrialisation, especially in view of the pre-emptive measures against Communist Party-ruled China by the Western bloc during the Cold War, a strategy still practised by the United States now. In other words, industrialisation was regarded as the vital means to securing independence and safeguarding sovereignty. Along this line of logic, the later 'open door' policy and marketisation, instead of representing a rupture with the developmentalism pursued by a late industrialising country, has in fact continued it. As long as the aim was development as rapid industrialisation, it was an essential question whether the means was collectivisation or the introduction of foreign capital. Therefore, once the shift in geopolitics provided the conditions, China opened its door to the capitalist world, by allowing access first to its labour resources and then to its domestic market.

According to Kong Xiangzhi's research, the contribution of peasants to nation building in the first 60 years of the People's Republic of China (PRC) was around ¥17.3 trillion, made possible by policies such as the price-scissors system of agricultural and non-agricultural products, the mobilisation of cheap labour and land acquisition (Kong and He, 2009).

Land: The most important stabilising factor in China

Despite this, the peasants were still willing to support the state's industrial policy, which was exploitative to peasant labour and land. This was partly because the Communist Party of China (CPC) had implemented and then completed land reform (1949–52).

CPC used the traditional slogan of 'land to the tillers' to mobilise hundreds of thousands of peasants to fight for land revolution and the national liberation movement.[3] After 1949, CPC came to power and implemented comprehensive land reform. Land was equally distributed among peasants. At least 85 per cent of the peasants enjoyed the benefits of land distribution. Each peasant household had, and most of them still have, a small parcel of arable land. The per capita arable land was 0.11 hectare in 2008. In other words, around 900 million small landowners are vastly dispersed throughout the whole nation.

China feeds 19 per cent of the world's population with only 8 per cent of the world's arable land (2011).[4] The total population has reached 1.3 billion.

According to the Ministry of Land and Resources of PRC, arable land is around 122 million hectares (2011),[5] about 13 per cent of the total area of the country. However, China's agricultural output is among the largest in the world. China's grain output has recorded growth for the eighth consecutive year. It reached 571.2 million tonnes in 2011, 140.5 million tonnes more than the output in 2003 (Wen, J. [no date]). Land distributed to the peasantry is utilised mainly for food production to maintain self-sufficiency. There are around 200 million small rural households and 680,000 villages. Each peasant household has an arable plot, which is ultimately under the direction of the village committee. In terms of legal entitlement, arable land is collectively owned by a rural community and distributed within the village according to the size of household and other factors. It is a form of collective ownership. As a whole, the majority of the population in China consists of smallholding (landowning) peasants.

Strictly speaking, the migrant (peasant) workers are not the proletariat, the classical definition of which being those who have nothing for the market except their labour power. The peasant workers have their own parcels of arable land for subsistence; they are not landless people. This is undoubtedly the legacy of the 1949 Revolution. One of its political achievements has been the realisation of material improvement for the majority of the people, i.e. the peasants. Nowadays, peasants and workers are increasingly suffering from exploitation and social injustice, but the legacy of land revolution, as well as a few residual socialist practices, still more or less insulates Chinese society from being ruthlessly plagued by neoliberal globalisation and its destructive projects of modernisation.

Since 1989 the contribution of agriculture to the GDP and peasants' household incomes has been declining. After 1993 the development of rural enterprises was systematically curbed in order to boost export-oriented growth (i.e. globalisation). This resulted in a massive flow of migrant workers from the rural areas into cities. These workers mostly consisted of the surplus labour force from rural households that owned a small arable plot. They were, therefore, different from the working class as defined by classical political economy, which derived from the expropriation of land. These migrant workers endured irregularly paid wages, accepted employment without social benefits and consciously suppressed consumption to collect (once a year in some cases) their cash income. What underpinned this practice has been a particular form of collective landownership. This has been the real foundation for China's ability to maintain low labour costs for 20 years. The rural sector has taken up the cost of social reproduction of labour, a cost that capital generally aims to shrug off. The so-called 'comparative advantage' theory is not enough to explain China's ascendency, because there was no shortage of developing countries with a huge population base (not to mention that a large surplus labour force could also turn into a source of social instability, which has not been the case in China).

The second important function of the rural sector is to serve as a buffer to absorb the institutional costs of the urban sector, which have been expressed as crises. In China, one of the crises repeatedly took the form of massive unemployment. There were three occasions before 1978 in which the regime initiated massive population migration to the rural areas through political movements. It was in fact a way to resolve the crisis of urban unemployment. After the reform, the rural sector has continued to stabilise Chinese society as a whole by two essential functions. Primarily, the rural sector continues to serve as a labour pool. But that alone cannot explain China's so-called 'comparative advantage' (abundant supply of cheap labour). Since unemployed labour in the urban sector can also result in social unrest, so-called advantage can turn into disadvantage.

The urban sector as a capital-intensive pool is necessarily vagarious and risk-generating, constantly destabilising the society through cyclic crises. On the contrary, the rural sector can regulate the labour market by reabsorbing unemployed migrant workers in from the cities in times of economic crisis. Its stabilising capacity lies in the rural land community ownership system that has remained intact to some extent even till today.

In China, land is not simply a production factor as simplistically theorised by mainstream economics. It also carries important social and cultural functions. As Karl Polanyi (1944) argues, land possesses qualities that are not expressed in the formal rationality of the market. During the 30 years since the reform, it has been an important factor in stabilising the society at large. In the rural sector, landownership is a form of collective ownership. Indoctrinated by the neoliberal ideology, many intellectuals in China nowadays advocate radical privatisation of land. Radical privatisation may facilitate and accelerate the commodification of land. But we must ask an essential question: who then takes a larger share of the institutional returns? Obviously it is not the smallholding peasant households with their last small parcel of land, but most likely the real-estate interest bloc and rent-seeking authorities. Who will eventually bear a greater part of the consequent institutional costs in terms of social destabilisation? Apparently, once again the powerless peasants. These problems are missing in the lopsided concept of efficiency/productivity as measured by gains in GDP growth through the commodification/monetisation of land. Non-monetised or non-monetisable factors like social stability and community integrity are essential to a society in development.

Land expropriation

Nevertheless, more and more peasants are losing their land. The government estimates that the current amount of arable land is roughly 122 million hectares, which remains unchanged since 2005. According to Tan Shuhao's research, the ratio of construction sites in arable land occupation has continuously increased from around 10 per cent in 2002 to 80 per cent in 2008.[6] The Ministry of Land and Resources disclosed that of the loss of arable land, 77 per cent goes to construction projects.

According to the *2011 China Urban Development Report* by the Chinese Academy of Social Sciences, the number of Chinese peasants who have totally or partially lost their land currently amounts to between 40 and 50 million. The number is going to increase by 2–3 million per year. Land expropriation is propelled by local governments and speculative financial capital. Since 2000, only 20–30 per cent of the capital gain obtained from value added to land has been distributed at the village level, and merely 5–10 per cent is eventually allotted to be shared by the peasants as compensation. Local governments take 20–30 per cent of the added value, whereas real-estate developers take the lion's share of 40–50 per cent. Out of the petitions filed by peasants 60 per cent arose due to land disputes. A third of these cases are related to land expropriation. Among those surveyed, 60 per cent are facing difficult living conditions, particularly with regard to issues of income, retirement and health care.

Local governments' fiscal constraint has been a major cause of extensive large-scale land expropriation. Since the reform, intermittent economic crises had confronted the central government in the form of deficit. The central government responded by adopting the policy of decentralisation of the tax and revenue system, which led to local governments' dependency on local revenues. In the period starting from 1984, local governments occupied farmlands for local industrialisation in order to generate income. It was the period of 'land for local industrialisation'. In 1994, China was confronted with a triple crisis (balance of payments, fiscal deficit and banking system). It was the year marking China's reckless embrace of globalisation. The central government implemented a drastic tax and revenue system reform. Before 1994 about 70 per cent of the local tax revenues went to local governments. But since then, about 50 per cent has gone to the central government. In order to compensate for the drop in the share of revenues local governments again appropriated farmlands to invest in commercial projects. This was the period of 'land for commercial fortunes'. Since 2003, local governments have increasingly collateralised farmlands for mortgage loans from commercialised banks. In the age of financialisation, it is the period of 'land for mortgage loans'.

Landless new generation

In 2003, the Law of the People's Republic of China on Land Contract in Rural Areas was promulgated. It stated that new inhabitants would obtain contracted land only if there was land reserved, land increased through reclamation or land turned back by other contractors. One possible consequence of this new legislation is to exclude those born since then from being beneficiaries of land distribution. Once arable land is no longer evenly distributed and the peasants no longer have an expectation to share in the benefits of land, the mechanism of risk management through internalisation in the rural community would be greatly weakened. The behaviour of migrant workers from rural regions as such is going to change quite fundamentally.

It is expected that the new generation of the rural population will radically dislocate themselves from agriculture and the rural regions. Nowadays, there are around 200 million peasant migrant workers in the cities. Unlike the former generations of migrant workers seeking employment in cities, the newer generations are no longer content with simply earning enough cash to maintain the reproduction of peasant households. Furthermore, cash income needed for expenditures like education and medical care have far exceeded that which can be afforded by localised labouring in agriculture. The will of the new rural generation to settle in the cities is in tandem with the government's policy of urbanisation. Moreover, they are no longer surplus labour from peasant households but, in essence, have finally evolved into the working class defined by classical theory. They are going to play an active role in the manifestation of structural contradictions of China's society during its transition. In view of these contradictions, the traditional agrarian sector may no longer serve as a reservoir of surplus labour as it used to under a dual urban–rural system. Therefore, the so-called 'comparative advantage' of China is being eroded.

Collective landownership in rural areas is an issue much neglected as the dominant ideology in Chinese intelligentsia and media is neoliberalism, respectively in its individualist and statist forms. At present, it is of the utmost importance that the legacy of the 1949 land revolution for small peasants be safeguarded.

Crisis: The Cost of Pro-Capital Reform and its Transfer to the Rural Sector

Wen Tiejun argues that between 1949 and 2009, China has undergone eight notable crises, and that the rural sector has always played the role of social stabiliser by absorbing the cost of crisis (Wen et al., 2012). The root of crisis has been the reckless pursuit of modernisation and industrialisation. The outbreaks of crises have been scattered along a trajectory marked by four instances of introducing foreign investment. The first of these occurred with the deterioration of China–USSR relations. Between 1950 and 1956 the USSR's total aid investment in China was worth US$5.4 billion. In 1960 the USSR aborted all aid and investment, thrusting China's economy into crisis first in 1960 and then again in 1968. The intensification of capital inevitably entails increasing risk. Introducing foreign capital in pursuit of industrialisation, whether the capital is Soviet or Western, makes a nation vulnerable to economic risk. Crisis is inexorably endogenous to capital.

The second instance of foreign investment playing havoc with China's economy began in 1971 when China accepted US$4.3 billion Western investment, leading to economic crises first in 1974 and then in 1979. The third instance occurred in the 1980s. Many local governments leapfrogged to attract FDI and therefore amassed a great deal of foreign debt, which again proved to aggravate economic crises, once in 1988, followed by another in 1993. All these economic crises can be regarded as internal crises derived from domestic fiscal deficits. China embraced globalisation in the mid 1990s, and the fourth instance of economic crises broke out in 1998 and 2008. These two crises can be categorised as 'imported crises' and were a consequence of the external financial crisis at the global level.

In the economic crisis of 1960, 12 million unemployed educated youths were sent to the rural areas in the name of receiving re-education by the peasants and building the new socialist village. In the crisis of 1968, another 17 million youths were sent to the countryside to release the pressure of large-scale unemployment. In 1974, more than 10 million youths were dispatched. The total number added up to around 40 million. By absorbing the unemployed labour force, the rural sector actually served to absorb the cost of crisis caused by the pursuit of modernisation. Wen Tiejun thus generalises a regularity of crisis and reform in China in the last 60 years. He concludes that if the economic crisis induced by introducing foreign investment could be contained by displacing the adverse conditions towards the rural sector and the crisis in the capital-intensive urban-industry sector could in this way be much abated, China would achieve a 'soft landing' and the existing institution could be maintained as the pressure is released. Otherwise, in the case of a 'hard landing' in the urban sector, the central government would be forced to initiate a 'reform' in the fiscal and economic system (Wen et al., 2012).

In reality, the so-called reforms, which were much hailed by the West as well as the official media and ideologues, were nothing more than a series of expedient measures in response to crisis, rather than being deliberately planned by wise leaders.

'Three-Dimensional Problem of Rural China'

The rural has been constantly appropriated and systematically exploited for national modernisation. It is in this context that Wen Tiejun coins the renowned notion of the 'three-dimensional problem of rural China' (*sannong wenti*). Wen explains that the problem of the rural sector in China cannot be simply regarded as an agricultural issue, but involves the interrelations between 'rural people (income disparity/migrant workers), rural society (multifold socio-economic issues and governance), and production (agricultural vertical integration/ township and village enterprises development)'. So by 'three dimensions' he means the peasantry, the villages and agriculture, none of which can be condensed into the other. It follows that China's rural problem cannot be solved simply by industrialising (modernising) agriculture according to the US model, as naively imagined by many advocates of modernisation. Although by 2012 the rate of urbanisation in China surpassed 50 per cent, about 600 million people still live in the rural areas. Even if we can set aside the unsustain-ability of industrial agriculture in terms of ecological devastation and energy consumption, the surplus labour force (maybe up to 200 million) thus liberated by highly mechanised agricultural production simply cannot be absorbed by the expansion of industrial capacity in the world.

In other words, peasant agriculture remains an indispensable mode of production in China, whether the single-minded advocates of modernisation like it or not. In the light of this, Wen Tiejun (2001) states that 'China's problem is the tension aroused in an agrarian society, characterised by overpopulation and limited resources, by the process of internal and primitive accumulation of capital for state industrialisation'.

'Rise' at the expense of the rural

In 2010, China stood as the second largest economy after the United States. According to IMF statistics, China's foreign reserves reached US$ 3.1 trillion in March 2011, which accounted for nearly one-third of the world's foreign reserves. According to the WTO secretariat, China's share of the global GDP was 9.6 per cent in 2008, 9.1 per cent in 2009 and 10.3 per cent in 2010. Nevertheless, this kind of 'rise' is achieved at a dear price. And among those who bear the costs disproportionately, the peasantry has shouldered the greatest burden.

As seen earlier in this chapter, at the initial stages of national modernisation the rural sector had been systematically exploited for accumulation. After China resumed diplomatic relations with the West and once again introduced foreign investments on a massive scale in the early 1970s, serious fiscal and debt crises broke out almost instantly. China's legendary reform and open policy in 1978 actually originated from a response to this crisis. After the implementation of the reform, peasants at first enjoyed the benefits of new policies and witnessed substantial improvements in income. However, in the early 1990s the central government systematically suppressed the development of township enterprises. The income growth of peasants has declined since then. The major turn took place in 1993, a year when China was struck by the triple crisis: fiscal deficit, balance of payments crisis and banking crisis. From then onwards China, in order to earn foreign exchange reserve to resolve the foreign debt crisis, suppressed the domestic market and embraced a predominantly export-oriented strategy, merging itself into globalisation. After almost 20 years of its participation in globalisation, China has now been facing the increasing pressure of global excess financial capital. The tension between domestic and international interests is approaching a critical point of explosion. However, the export-oriented model has become such a deep-rooted path dependency that China has to make a great effort to switch its trajectory of development.

Despite the stunning economic growth, the environmental and ecological devastation is cataclysmic. Water and air pollution is constantly at harmful levels. Of the world's 20 most air-polluted cities 16 are located in China, with a population of 400 million living under daily threat. One-third of the land is contaminated by acid rain and almost 100 per cent of the soil crust is hardened. China has become a dumping ground of waste from the West. Waste is one of the top three US export 'goods' to China and the one with the fastest growth.

The National Bureau of Statistics announced that according to the sample survey and comprehensive statistics conducted in 31 provinces throughout the nation, in 2010, the total grain production was 54,641 million tonnes, which was an increase of 1,559 million tonnes, or 2.9 per cent, when compared with 2009 (NBS, various years). This is the seventh consecutive year of increased grain production. However, at the same time, the use of chemical fertilisers has increased from around 1 million tonnes in 1979 to around 5.5 million tonnes in 2009. Industrial agriculture has become the largest source of water and soil pollution in China. And it is the peasantry who suffers most from chronic agrochemicals poisoning.

According to China's State Environmental Protection Agency, in 2006, 60 per cent of the country's rivers were too polluted to be sources of drinking water. Continuous polluted emissions come from industrial and municipal sources, as well as from pesticides and fertilisers (SEPA, 2006). This crisis is compounded by the perennial problem of water shortages, with 400 out of 600 surveyed Chinese cities reportedly short of drinking water. According to the Ministry of Water Resources, roughly 300 million people, most of them rural residents, do not have access to safe drinking water.

The social cost of specialising in low-end manufacture is also enormous. In China it is estimated that nearly 200 million people suffer from occupational diseases; over 90 per cent of them are migrant workers from rural areas. In the Pearl Delta Zone alone, each year at least 30,000 cases of machinery-induced finger-cut accidents are reported, with over 40,000 fingers mutilated. Again, most of the victims are migrant workers from the rural areas (70.2 per cent; merely 4.3 per cent are from the cities) and many of them fail to receive any compensation in the end (Zhang, 2005).

At present, China is facing three major structural contradictions. The first is the huge income gap between the urban and rural sectors; and the second is the developmental disparity between the coastal regions and the hinterlands.

The peasantry is directly bound up in these two contradictions. The third is the conflict in development strategies between industrial and financial capitals. The former, confronted with excess capacity and fierce international competition (therefore a declining marginal profitability), will become even more vulnerable as the financial sector (largely state-owned monopoly capital) pushes forward monetary liberalisation in order to take a greater part in global financial capitalism. Interestingly, in the initial stage of globalisation, the rural sector was sacrificed for the industrial sector. Now in the stage of financialisation, the industrial sector may be in turn sacrificed for the interests of financial capital.

Raw money power

Being pro-capital is often a policy proclivity when a nation pursues indus-trialisation under conditions of capital scarcity. This has profoundly shaped the governmental behaviour in emerging countries. One of the institutional contradictions in contemporary China is the disparity between the central government and local governments. The central government pursues state capitalism and takes firm control of various monopoly capitals, whereas local governments are modelled by government corporatism. Local governments at different levels become increasingly rent-seeking. The central government with a handsome fiscal surplus can afford to orient itself more towards pro-poor and pro-people livelihood policy. However, local governments at various levels under budget constraints remain highly pro-capital. This structural imbalance has become an institutional contradiction affecting China's policy viability.

Since 2003, the Chinese government has started to focus on solving rural problems. A series of pro-rural poor policies have been implemented: the elimination of agricultural tax, comprehensive aid to agriculture, the cooperative medical service system, the cancellation of educational fees in poor western regions, a substantial increase of governmental investment in public services, and new rural finance policies, among others.

In October 2005, the Chinese government highlighted the 'new rural development' as a national strategy. The Central Government's No.1 Document, issued in February 2006, illustrated that 'the building of a new socialist countryside' is 'characterised by enhanced productivity, higher living standards, healthy rural culture, neat and clean villages and democratic administration'. Meanwhile, Hu Jintao, general secretary of the Central Committee of CPC, emphasised: 'As the resolution of issues concerning agriculture, rural areas and peasants [*sannong wenti*] has an overall impact on China's target of building a moderately prosperous society, in all respects, we must always make it a top priority in the work of the whole Party.' In October 2007, the articulation of an 'ecological civilisation' was set as a guiding principle.

According to the statistics, from 2004 to 2010, the central government increased its investment in the rural sector to ¥857.97 billion. The annual rate of increase is 21.8 per cent. The investment for grain production has increased from 102.9 billion to 457.5 billion.

In the last decade, the investment in rural society has enabled China to tackle the external crisis. For example, in 2008 when the global financial crisis broke out, 20 million peasant workers in the coastal areas lost their jobs. A sudden upsurge of unemployment on such a large scale would mean social and political disaster in any country in the world. Yet, no major social unrest happened in China. The peasant workers simply returned to their home villages to sit through the period of temporary unemployment. It was because they still had a small plot of land, a house and family to rely on as a last resort. In other words, the smallholding in the village is a peasant worker's 'base of social security'.

Apart from the efforts by the government at various levels to solve the rural problems, some villages have negotiated with the forces of modernisation, marketisation, urbanisation, atomisation and monetisation of social relations, which are destroying rural society.[7]

As David Harvey points out, with the advent of capitalism, 'money was the power of all powers', referring to the raw money power that dissolves the traditional community. He further elaborates:

So we move from a world in which 'community' is defined in terms of structures of interpersonal social relations to a world where the community of money prevails. Money used as social power leads to the creation of large landed estates, large sheep-farming enterprises and the like, at the same time as commodity exchange proliferates. (Harvey, 2010: 294)

In an attempt to assert its authority of governance or reverse the degradation of the rural society, the central government, along with village committees, has endeavoured to address the detrimental role money plays in destroying social relations. However, the focus of its solutions (such as increasing investment in rural areas or sharing profit equally) is still in terms of money. In that sense, the government is not critical of the destructive aspects of modernisation or developmentalism.

An Alternative Path: China's Rural Regeneration Movement

Today, the rural reconstruction movement is the biggest social movement in China, with tens of thousands of volunteers, yet peaceful (Wen et al., 2012). It traces its intellectual lineage to the rural reconstruction movements before the Japanese invasion in the 1930s.

Capitalism invaded China soon after the First Opium War of 1840–42. The traditional social order started to disintegrate and crumble. However, an integration of peasant agriculture, household industry and village community was resistant to historical change: this was what Marx referred to as the Asiatic mode of production. The notion ignited a debate among Chinese intellectuals about China's history and future.

The 'peasantry' was considered the stagnant and backward element that had become a hindrance to China's progress. Both rightist and leftist intellectuals largely embraced the idea of 'modernisation' in the name of 'science' and 'democracy'. It was believed that China should pursue industrialisation in order to resist imperialist invasion. However, there was another intellectual trajectory critical of industrial modernisation, which took the small peasantry as the starting point and base for China's transformation.

Some famous modern Chinese intellectuals, such as Liang Qichao (1873–1929) and Liang Shuming (1893–1988), challenged Marx's idea of the five stages of world history, namely primitive communism, slavery, feudalism, capitalism, and socialism or communism, arguing that China's nature included a kind of rural governance based on small peasantry and village community, and a combination of private and public ownership of land and labour. This kind of rural governance had existed for at least 2,000 years. In other words, they objected to the imposition of Marx's idea of the linear development of world history on China, but they agreed with his diagnosis of Chinese society as having the characteristics of an Asiatic mode of production.

Marx admitted that Asia was beyond his knowledge. Through reading books, reports and other materials written by colonialists at that time, Marx articulated that the Asiatic mode of production was mainly based on 'the unity of small-scale agriculture and home industry', and 'the form of village communities built upon the common ownership of land'.

Claude Lefort considers that according to Marx the Asiatic mode of production is generally based on the double determination of the individual, as a property owner and a member of the community. Each individual has the status of proprietor or possessor only as a member of the community. Communality of blood, language and customs are the primordial conditions of all appropriation (Lefort, 1986: ch.5). In his *Grundrisse*, Marx remarked that 'land is the great workshop, the arsenal which furnishes both means and material of labour, as well as the seat, the base of the community' (Marx, 1973: 472).

Therefore, Marx elaborates,

In the oriental form the loss [of property] is hardly possible, except by means of altogether external influences, since the individual member of the commune never enters into the relation of freedom towards it in which he could lose his (objective, economic) bond with it. He is rooted to the spot, ingrown. This also has to do with the combination of manufacture and agriculture, of town (village) and countryside. (Ibid.: 494)

As Lefort further elaborates, the communes are sheltered from all the torments of the political domain, but also a given mode of communal existence proves to be shielded from outside attacks. And this simplicity has made Asiatic societies endure social stability. Marx later remarks:

> The simplicity of the productive organism in these self-sufficing communities which constantly reproduce themselves in the same form and, when accidentally destroyed, spring up again on the same spot and with the same name – this simplicity supplies the key to the riddle of the unchangeability of Asiatic societies, which is in such striking contrast with the constant dissolution and refounding of Asiatic states, and their never-ceasing changes of dynasty. The structure of the fundamental economic elements of society remains untouched by the storms which blow up in the cloudy regions of politics. (Marx, 1976: 479)

Although the idea of a changeless Asia not affected by the general progress of history is a Eurocentric fabrication, Marx did capture some aspects of the foundation of the social stability in Asia. The tenacious capacity for recovery of China's rural society lay in internal cooperation and the management of common resources.

Liang Qichao, a renowned modern intellectual and politician, visited Europe during 1918 and 1919. He had been involved in pushing for Western democracy and parliamentary government, but changed his views completely after witnessing the war and the disaster in Europe. He went back to studying Chinese traditions. In *A History of Chinese Culture* (1923), he concluded that Europe was based on urban governance, whereas 'China is based on village governance but not urban governance'. Village governance is composed of two main factors: small peasantry and village community. He argued that small peasantry has been the nature of China's society for at least 2,000 years; it is derived from the practice of dividing up property among family members. He further elaborated that during the Qing Dynasty (1644–1911), it was legally required that family property should be divided up equally among the offspring. In that sense, the bulk of them were smallholding peasants.

The majority of the Chinese population settled along two main rivers, the Yellow River and the Yangtze River. A single village or a peasant household could not individually solve the problems of irrigation, such as flood and drought. The imperative of survival required a cluster of villages along the rivers to work together to manage public affairs and to deal with external crises. So the major concerns were about an arrangement of cooperative collective labour and the protection of common property. Local governance was derived from village community building that paved the way for the development of nation building. Chinese civilisation has been based on irrigation, small-scale agriculture, small peasantry and village communities.

Moreover, village communities usually contain three crossed layers of relations: kinship (blood), neighbourhood (locality) and agricultural fellows (peasants). Village communities not only solve the external crisis, such as natural disasters, but also turn the crisis into the reinforcement of the capacity of crisis management. This, nevertheless, requires mass mobilisation among peasant families and village communities. Thus, the practice of sharing common property as well as solving common problems is inclusive and cooperative.

During the 1920s the rural reconstruction movement attempted to reactivate the Chinese tradition of small-scale agriculture and home industry. Liang Shuming (1893–1988) was one of the leaders of the movement. He was not only a Confucian and Buddhist intellectual but also a political and social activist. He was involved in the reconciliation between Kuomintang and the Chinese Communist Party during the Sino-Japanese War (1939–45). In 1977, he reflected on his engagement in the rural reconstruction movement during Republican China: 'In the very beginning, I childishly believed that we must learn from the West. Shortly afterwards, I realised that it was impossible for China to become a westernised capitalist society. So, I had the idea of the village as the national base' (Liang, 1977: 424–28).

In 1937, Japan, an emerging capitalist country, invaded China. Liang Shuming was forced to stop his experiments of rural construction. In the same year, his book *Theory of Rural Reconstruction* (also entitled *The Future of the Chinese Nation*) was published, in which he theorised his working experiences in the Institute of Village Governance in Henan Province in central China (1929–30) and the Research Institute of Rural Construction in Zhouping Township of Shandong Province in north China (1931–37). Counteracting Western and Japanese imperialism and going against the dominant understanding, Liang did not urge for complete westernisation and industrialisation in the way that Japan did. He not only condemned foreign imperialists but also reprimanded Chinese nationalists and radical revolutionaries, as he believed that they were fundamentally destroying rural society. Although Liang was born into an urban intellectual family, he considered the rural areas as the foundation of Chinese rule and democracy. He proclaimed:

The foundation and the centre of Chinese society is the village. All cultures mainly come from and are practised in rural society – for example, the legal system, secular customs and commerce, among others. Over the past hundred years, imperialist invasion certainly destroyed the countryside, directly and indirectly. Even the Chinese people ruined the village, like those revolutionaries who were involved in the Hundred Days Reform or the nationalists who promoted national self-salvation. Therefore, Chinese history over the past hundred years is also a history of village destruction. (Liang, 2006: 10–11)

In the face of village destruction, Liang devoted himself to the rural construction movement. Liang's experiments included 'the village school as the basic administrative unit', organisation of peasants' associations, setting up of cooperatives and small-scale village industries, and improvement of agricultural technologies, among others.

Liang designed the village school as a learning unit that comprised local elites, common villagers, and outsiders including intellectuals and professionals. The aim was to activate the communal capacity of problem solving at the grassroots level. Therefore, Liang's theorisation of and praxis for the future of China is rooted in the village community. He treats 'the rural' as an alternative to modern capitalist society.

Liang mentioned that village regeneration is the means of the revival of Chinese culture. Rather than being a conservative and chauvinist Confucian, Liang reinforced the importance of nurturing 'new ethics' from the Chinese tradition, which could make one differentiate oneself from the aggressive bourgeois culture and belief. He criticised the facts that the powerful development of Western culture was based on a drive 'to conquer Nature and to take advantage of Nature', and that capitalism is 'individualistic and self-centred'.

Liang used a metaphor of 'new buds on the old tree' to describe the rural reconstruction movement. In 1977, he wrote a paper to reflect on his experiences of rural reconstruction, in which he concluded that rural reconstruction was a question of ethics, '[t]o be positive towards life and to remember the importance of ethics and friendship', which was a challenge to the capitalist value system. Furthermore, he explained revival of the 'Chinese culture':

If you ask me, 'what is actually the revival of Chinese culture in the world in the near future?' I will simply answer that when it proceeds from socialism to communism, religion declines and is replaced with a self-awakening and self-disciplined morality; national law disappears and is replaced with social customs. (Liang, 1977)

Another famous leader of the rural reconstruction movement is James Yen (1890–1990). Yen dedicated his life to the education of the *ping-min* (the common people). He served Chinese coolies working with the Allies in France during the First World War. In particular, he helped the illiterate coolies to write letters to their families in China. This experience of working with the poor enabled him to promote the literacy campaign. After returning to China, Yen organised mass education and was involved in the rural reconstruction movement in 1923. The 'ping' (literally meaning common, ordinary and equal) was the logo of the mass education and rural reconstruction movement founded in China in 1923, and is also the logo of the International Institute of Rural Reconstruction (IIRR) initiated in 1960.

Yen thought that the majority of the poor were rural people plagued by poverty, physical weakness, ignorance and selfishness. So, it was necessary to improve the quality of peasant life and then of rural society. Yen also saw the basis for a new Chinese nation in rural reconstruction. The area where he conducted his experiment was Ding County in Hebei Province, some 322 km south of Beijing. Working together with the village committee and local government, Yen coordinated innovations ranging from hybrid pigs and economic cooperatives to village theatre and health centres. His work was disrupted by the Japanese invasion of 1937. He later founded the IIRR in the Philippines in 1960.

Following Liang's and Yen's spirit of rural regeneration, a new rural reconstruction movement emerged at the turn of the twenty-first century. Its background has been rural degradation while China's export-led manufacturing industries and the demand for cheap labour are besieged with a world economy battered by financial crises. There has been a heated debate about the *sannong wenti* (three dimensional aspects of the agrarian issue) in the academia and media. Against this background, some intellectuals, NGO workers and local villagers have worked together to explore ways of regenerating rural society, with some viewing it as part of their poverty alleviation work, and others seeing their commitment as providing another mode of modernisation, in the spirit of Liang and Yen, different from the mode of development of the West (urbanisation). The first initiative was the James Yen Rural Reconstruction Institute (2004–2007), which provided peasants with free training courses and mobilised university students to work for the villages. Apart from that, Green Ground Eco-Center was founded in 2006, promoting ecological farming and rural–urban cooperation. Little Donkey Farm was established in 2008, with an area of 230 *mu* (about 15.3 hectares) and situated in the suburbs of Beijing; this is a partnership project between Haidian District Government and Renmin University of China. It promotes community-supported agriculture and facilitates rural–urban interactions. The Liang Shuming Rural Reconstruction Centre was set up in 2004, to provide university students with training programmes for working in the countryside.

These experiments are based on the following perspective: with the advent of capitalist modernisation and developmentalism, raw money power has caused the gradual deterioration of rural society and communal relations. The solution usually adopted by the government or village committee is one that revolves around the increase of money investment. Hence, cash investment and profit-sharing are typical measures. But human relations to the land and the community, largely damaged by modernisation, are yet to be addressed. In other words, the ultimate concern must be how to rebuild one's ties to nature and to others. Peasant agriculture is an important way of repairing human relations to Mother Earth. Currently, the food system of the world is mainly controlled by the capitalist transnational agro-companies, which make huge profits through mechanised and chemical monoculture. Countering this trend, peasant agriculture and small peasantry practising organic farming and having local knowledge should be protected and promoted. In this way, organic food products can be one of the foundations of rural–urban solidarity. At the same time, communal capacity should be activated in terms of the utilisation of common resources and participation in the problem-solving process. This requires cooperation between grassroots people and intellectuals.

Another example of rural regeneration is the Yongji Peasants' Association of Shanxi Province. It was formerly the Center for Women's Cultural Activities and Women's Association, established in 2003. Now it has 3,865 members from 35 villages in two counties. It includes six technological services centres, a handicrafts cooperative, steamed buns workshops and an ecological agriculture zone. Socialised voluntary labour, redistribution of resources and concern for the younger generation are central to these initiatives.

The feeling of solidarity that arises from participation in collective activities rooted in daily practices can be life transforming, embodying Marx's conception of revolutionary practice as a conjuncture of social change and self-change. By devoting labour to social redistribution rather than capitalist accumulation, peasants take pleasure in helping others as they gain others' respect for their contributions. Working for others through socialised labour may mistakenly be regarded as a residual practice in a rural society, but it is also radical practice in the face of the forces of globalisation and the hegemonic mentality of individualism and entrepreneurship. Building a culture of collectivity through daily practices of voluntary labour and redistribution of profits is a profound mode of being that counters the violence of capitalist economic endeavours.

Rural regeneration and new historical agency

> *Who controls the food supply controls the people;*
> *who controls the energy can control whole continents;*
> *who controls money can control the world.*
>
> Henry Kissinger

At this point we must ponder a pressing question: what is the specific historicity at present that accentuates the historical agency of rural regeneration nowadays?

Three decades of globalisation have shown the reckless ascent of unfettered financial capitalism. In its present stage, globalised financial capitalism is centred around currency hegemony. The Bretton Woods regime has set up the

US dollar as the dominant global currency. After the abandoning of the gold standard in 1971, the dollar has been given a free reign to increase money supply without limits to the world, while the United States enjoys a form of seigniorage as the dollar is set as the major settlement and reserve currency in the world. Oil has become geopolitically vital as it serves as a new base to secure the dollar's value. Financial products add to the list of vital commodities, as a majority of the world's financial products are valued in dollars. And the most important pillar of the dollar's hegemony is US military power. It is no wonder that US military expenditure alone accounts for 50 per cent of the total amount of money in the world. In place of the industrial–military complex, now there is the geopolitically pervasive, omnipresent financial–military complex. In this sense, the overarching shaping force of the world order is no longer geopolitics but currency politics. Geopolitical presence becomes less of a determining factor than the hegemonic presence of the dollar in a currency zone.

It is hard to imagine a better way to do business than exchanging physical commodities with pieces of printed green paper. The only setback is the nominal liability of public debts. This is no problem – as long as the United States remains the mightiest military power in the world! The debts' issue can be partially resolved by continuously injecting money into the system. Since the financial crisis in 2008, the United States has been dumping trillions of dollars into the world market as a strategy to dilute its debts and hence transfer its cost of financialisation to the world. As a result, the prices of major commodities, most importantly agricultural products and oil, are going through the ceiling. Finally it has become apparent why the United States and the European Union are so keen to protect their own agriculture while disarming most of the other nations' food sovereignty. No wonder agriculture has always been the key issue in WTO negotiations.

The theory of 'comparative advantage' has it that if you can buy cheap food from abroad, why bother growing it yourself? Grow cash crops instead, or 'upgrade' your economy from a backward primary industry to a secondary one, but be content with low-end manufacture as cheap labour is your 'comparative advantage'.

However, the age of cheap crops has gone. By controlling oil one controls the modern industrial system, whereas controlling food supply is the way to subject the people to the yoke. Without petroleum there is no modern civilisation. But without food (and water) there is no civilisation at all.

Now geopolitical tension is less about regional presence or direct control than about a strategy of currency politics. For example, conflicts and wars in oil-producing regions are not so much about direct control of the oil supply as about maintaining high oil prices to absorb the expanded money supply.[8] Likewise, agro-fuels will never solve the problems, as has been claimed; on the contrary, they will produce more and greater problems (Houtart, 2009).

Agro-fuels are promoted because they push up global prices of crops and exert tighter control on food supply. Food production, no less than food supply, is one of the focal points of the new strategy of currency politics. Industrial monocrop agriculture is situated at multifold strategic points in the capitalist dominance and realisation of profits.

It is against this new historicity that rural regeneration, with the peasantry as one of the subjects, effectuates a new historical agency.

Capitalism must be transcended for our civilisation to be sustainable, and indeed to be civilised at all. But we must not naively believe that capitalism has exhausted all its possibilities. Otherwise we would be no less ridiculous than the liberalist 'end of history' ideologue. Capitalism never functions as neatly as its liberalist apologists or Marxist critics theorise. In addition to its capacity to constantly innovate, the vitality of capitalism consists of its monstrous ability to articulate different kinds of mode of production, including pre-capitalist modes, and subjugate them to the capitalist system. The origin of capitalism is flagrant enslavement and plunder. Marx is well aware of this as he denounces the myth of capitalist accumulation, the illusion of the immanent self-reproduction of capital. He says: 'In times long gone by there were two sorts of people; one, the diligent, intelligent, and, above all, frugal elite; the other, lazy rascals, spending their substance, and more, in riotous living' (Marx, 1976: 873). So, interests and capital gains are justified by the capital owner's willingness to suppress instant consumption. Further, Marx says: 'In actual history it is notorious that conquest, enslavement, robbery, murder, briefly force, play the great part... As a matter of fact, the methods of primitive accumulation are anything but idyllic' (ibid.: 874). So, Marx presents the famous notion of primitive accumulation, which precedes capitalist accumulation; an accumulation which is not the result of the capitalist mode of production but its point of departure. However, he does not stop theorising an immanent mechanism of the reproduction of capital, which would suppress and negate all other modes of production, encompassing all of humankind, and create the endogenous condition for its abolition.

But the trajectory of capitalism has not revealed itself in this way.[9] Global capitalism is an antagonistic system that articulates other heterogeneous modes of existence. Even nowadays slave labour in Brazil fits seamlessly within the country's industrial agriculture and thus feeds global capitalism. And we must say that capitalism is a total enslavement of nature and of other species. The brutal primitive accumulation is never merely a prelude to the capitalist mode but rather always its very foundation, in view of the world capitalist system. In this light, neoliberalism, with its ruthless expropriation of the global common, is an atavism. It may be said that capitalism can function only by maintaining a subtle boundary between the capitalist mode and others. Capitalism is global but never universal. The core capitalist nations can resolve the endogenous internal antagonisms only by transferring the cost to the outside. Therefore, the capitalist system is essentially heterogeneous, hostile and incessantly renovating itself, even through self-destruction. That is exactly what we are afraid of. Capitalism with its financial–military complex is bound to be even more vicious, violent and anti-civilisation.

Rural regeneration situated at one of the focal points of contemporary struggle, therefore, emerges with new historical agency. The overcoming of capitalism is an urgent historical project. But it is an open project. It calls for rethinking modernisation in order to open up the horizon and possibility of history again. Modernisation as a historical project becomes a linear and single trajectory, equivalent to industrialisation or the march toward capitalism. But whenever someone dictates a linear and single totalising path to us, we have every reason to be suspicious of a scheme in the service of partial interests. As Latour (1993) suggests, the myth of modernisation involves a 'purification' of temporality. The present is viewed as purely modern, distinct from a past that is outmoded and ineffective, and separates us from our benighted ancestors. We should rethink the distinctions between nature and society, human and thing, the past and the present, the rural and the urban, and between ourselves and our ancestors.

That is why the Zapatista insurgency effectuates such a strong historical agency. It rebels against the long-lasting monstrous repercussion of 500 years of capitalist history. It subverts all the distinctions between pre-modern and modern (and even postmodern), non-capitalist and capitalist, etc. When articulating a full spectrum of particular and singular struggles (race, gender, culture, territory, community, language, post-colonial, self-governance, etc.), it is not universal chez Hegel–Marx, but total (Ceceña, 2004).

An important form of historical violence is attributing a lack of agency to the dominated (subalterns) and excluded groups. For the secret of capitalism is often silenced exclusion rather than exploitation (wage labour). The ecosphere and other species are excluded from having a non-anthropocentric intrinsic value to exist in themselves, indigenous people are excluded as subhumans, and peasants as second-class citizens. This is so not because capitalism is not well developed in these realms but because the very exclusionary mechanism is endogenous to it. Therefore, to overcome capitalism at this historical conjuncture, a challenge is to re-effectuate the agency of these groups who have previously been stigmatised with lack of agency (portrayed by Hrabal as 'people abandoned in the rubbish heap of history').[10] The ecosphere and the peasantry are among the most important.

The historical agency of rural regeneration entails open potentiality and efficacy. We cannot discuss it at length in this chapter. Rather, we will highlight here the community and the common.

One of the central capitalist processes is dismantling the common by expropriation (plunder, privatisation or nationalisation) or mediation (for example, credit creation by banks). In place of the dismantled common, imaginary collectives ('civil society', 'the state', 'the race', etc.) must be set up. One of the conditions that make rural regeneration a valuable initiative in the historical cause of overcoming capitalism is the fact that in the rural community a rich heritage of the common is usually still available.[11]

It is well known that capitalism (aka modernisation) proceeds side by side with an inevitable breaking up of the 'restricted relationships' of all kinds ('all that is solid melts into air' [Marx and Engels, 1882]), most predominantly between (wo)man and land (nature), as well as among human beings. The breaking up of bondage of all kinds is regarded as an indispensable condition of historical progress. Liberalism thus mythologises an atomised individual at its ideological core. These individuals (often modelled in the image of high-income middle classes in capitalist metropolis) are bound up by nothing other than private property relationship. (Interestingly, Marx's proletarian as deprived individual is ontologically the former's mirror image.) However, private property is a myth. So-called private property is actually a specially managed form of the common. For example, money as the prime private property must first of all function as a social tool. Capitalist private property relationship is actually a subtly covert appropriation of commonwealth to serve the interests of special social groups. An atomised sense of existence is instrumental both in covering up the appropriation of the common and consolidating representative democracy, which has degraded into a defensive mechanism of the status quo by immobilising people's political and historical agency.

Paradoxically, only a pack of individuated social beings require the passive representation of a 'general will' by an avant-garde party or a partisan political organisation. This is because an active political will (or a historical consciousness) can form only when the common is experienced. The capitalist blocs, especially the financiers nowadays, are the only social groups that have an effective political will and historical agency, because only they have a clear vision of their appropriation of the common. People, reduced to atomised beings, are blind to the common they are deprived of.

To overcome capitalism, then, at issue with 'the masses resulting from the drastic dissolution of society' (*Critique of Hegel's Philosophy*) is the formation of people's agency through reconnectivity. The idea that people have to go deep into the capitalist relationship in order to transcend capitalism is of course very Eurocentric.[12] If, as mentioned, the tenacity of capitalism lies in its capacity to articulate with non-capitalist modes of production, then we cannot see why we should not articulate with what is valuable in non-capitalist modes in order to transcend capitalism.

Hardt and Negri (2009) describe how Marx in his old age loosened his progressivist stance. On one occasion he was asked to adjudicate between two groups of Russian revolutionaries: one side, citing Marx's own work, insists that capitalism has to be developed in Russia before the struggle for communism can begin; and the other side sees in the *mir*, the Russian peasant commune, an already existing basis for communism ... 'If revolution comes at the opportune moment,' Marx writes, '*if it concentrates all its forces so as to allow the rural commune full scope, the latter will soon develop as an element of regeneration in Russian society and an element of superiority over the countries enslaved by the capitalist system*.' (Quoted in Hardt and Negri, 2009: 88–89; emphasis added)[13]

We believe this is exactly what rural regeneration is all about – overcoming capitalism by rediscovering these valuable elements, such as the practices of cooperative labour (creativity), collective ownership (sustainable management of the common) and communal credit creation.

Of course, it must be emphasised that rural regeneration is not simply harking back to the traditional forms of rural community or nostalgia for an idyllic past. In fact, the parochialism of the traditional rural community must be fully recognised and transcended. But it can be achieved only through a patient and gradual transformation. External agents could humbly facilitate the process, but they should be cautious of any missionary or avant-garde mentality. The rural regeneration movement should be supplemented with expanded awareness, such as gender, eco-justice and good governance. In this way, instead of the Hegelian *aufhebung* to civil society and the state, the rural community can remain rooted in its localised finite form and yet transcend itself towards a richer agency.

Claude Lefort asks an astounding yet most meaningful question about Marx's thought: 'Should we say that [the proletariat] is the destroyer of the social imaginary or the last product of Marx's imagination?' (Lefort, 1986: 180). Maybe the peasantry with its historical agency, not unlike the proletariat, is a social imaginary, too. But it is a timely and efficacious one.

Concluding Remarks

Since the late Qing Dynasty, regardless of ideological preferences, Chinese intelligentsia and politicians have uncritically adopted the models of industrial and, later, financial capitalism at the expense of the peasants, the majority of China's population. This has led to the three-dimensional rural issues of peasant, village and agriculture. If 'rural China', or rural governance based on small peasantry and village community, is sustained for the cultivation of interdependent and cooperative relations within a community and among neighbouring communities, not only does it protect the livelihoods of the majority of the population but it also functions as 'a resistance' to the external crisis derived from global capitalism. In that sense, the current official experiments of building socialist rural areas as well as the activists of the rural reconstruction movement are contributing to the defence and justification of the small peasantry and village community, amid the disasters induced by capitalism. In summary, China's ascent is based on the exploitation of rural China. But the continuous experiments of rural reconstruction may provide an alternative to destructive modernisation.

Notes

1. The authors would like to thank Professor Wen Tiejun for his invaluable advice and Kho Tungyi for his help.
2. *Yang Wu* literally means 'affairs related to the West'.
3. Mao Zedong had rejected the orthodoxies of the CPC, then under the leadership of the Third International initiated by Moscow. He stood against the Stalinist doctrine adopted by the CPC leaders, which prioritised the industrial proletariat in cities as the revolutionary class. For a nation with over 80 per cent of the population as peasants, the orthodoxy was out of touch with realities. Mao legitimised the peasantry as the revolutionary class and emphasised land redistribution as the basis of forming revolutionary will. In a time when internationalism was manipulated by the USSR in service of its geopolitical strategy, Mao did not shy away from using nationalism to invoke guerrilla warfare against Japanese imperialism. It was the emphasis of the role of the peasantry in the making of a nationalist revolution that made Maoism the most predominant ideology adopted by anti-colonial as well as peasant guerrilla movements in the twentieth century. Che Guevara has affirmed that the guerrilla fighter is above all an agrarian revolutionary. Nevertheless, after the revolution Mao supported the Soviet model in order to force through an accelerated industrialisation at the expense of the peasants. And after breaking up with the USSR, the Soviet model thus built had become a path dependency, the bureaucracy a privileged ruling class without any equivalent economic base. Mao, complicated revolutionist and nationalist as he was, then swung back to the radical pole, invoking popular revolt in hopes of overthrowing the bureaucratic class, which came to be known as the Cultural Revolution.
4. Sources: Food and Agriculture Organization of the United Nations; United Nations Population Division and World Population Prospects.
5. Source: www.mlr.gov.cn
6. Tan Shuhao in a presentation at the International Conference on Comparative Studies for Sustainable Development, Renmin University of China, Beijing, 9–10 July 2011.
7. Huojiagou Village Enterprise of Shanxi Province is an example of practising the values of equality and solidarity when faced with the forces of individualism and monetisation. The village community covers 5 sq km, with 191 households and a population of 776. A small coal mine was the primordial resource for Huojiagou's industrialisation. Later, they invested in building a refinery and a power plant. The village demonstrated equality and solidarity through the fair distribution of wealth. For example, in December 2004, the assets of the enterprise were about ¥500 million.

The net assets were worth ¥300 million, of which 33 per cent was reserved for the village community. The remaining 67 per cent became shares distributed to the villagers, in three parts: individual share, seniority share, and post and duty share. They still insist on collective ownership despite intensive capitalisation.

8. This idea is inspired by Professor Wen Tiejun.
9. The Marxian history of primitive communism from slavery to feudalism, capitalism, then to socialism and finally communism is too linear to fit the real progression of capitalism. The Marxist historical notion is still bound up with the imaginative horizon of Eurocentrism (Young, 2004) with its peripheral blind spot to the colonised and peripheral world. The relationship of production does not always develop forwards. It often harks backwards in order to achieve higher productivity (higher exploitation rate). Instead of the linear history as portrayed by the West, the history of capitalism is often warped.
10. In both rightist and leftist theories, people have to get involved in the capitalist system in order to secure a place in historical progress. For Marx, only the working class has class consciousness, i.e. historical consciousness. Only the proletariat could exist as a historical agent. For those who are excluded from rather than exploited by the capitalist system, there is no historical agency. When criticising capitalism, Marx is most capitalistic.
11. Michael Hardt and Antonio Negri's *Commonwealth* (2009) contextualises itself mainly in the metropolis of core capitalist nations. The authors touch on the periphery in discussing the notion of altermodernity.
12. Recall Marx's early notorious view that colonisation was necessary for progress since it introduced the colony to capitalist relations of production ('The British Rule in India' and 'The Future Result of British Rule in India').
13. Later, in the preface to the Russian version of *The Communist Manifesto*, Marx and Engels (1882) write:

> The Communist Manifesto had, as its object, the proclamation of the inevitable impending dissolution of modern bourgeois property. But in Russia we find, face-to-face with the rapidly flowering capitalist swindle and bourgeois property, just beginning to develop, more than half the land owned in common by the peasants. Now the question is: can the Russian *obshchina*, though greatly undermined, yet a form of primeval common ownership of land, pass directly to the higher form of Communist common ownership? Or, on the contrary, must it first pass through the same process of dissolution such as constitutes the historical evolution of the West? The only answer to that possible today is this: If the Russian Revolution becomes the signal for a proletarian revolution in the West, so that both complement each other, the present Russian common ownership of land may serve as the starting point for a communist development.

5

ASIA (II)

The Political-Economic Context of the Peasant
Struggles for Livelihood Security and Land in India

Utsa Patnaik

The peasantry and rural workers of the global South are under historically unprecedented pressures today with respect to attacks by capital not merely on their livelihoods but on the very means of securing those livelihoods, namely the land they possess. Recalling the primitive accumulation of capital, which marked the birth and adolescence of capitalist production in Europe from the sixteenth to the nineteenth centuries, we see once more, albeit in different forms and under different circumstances, a concerted attempt by global capital and local corporations, on the one hand to acquire control over the *use* of peasant lands to serve their own purposes, and on the other, to seize that agricultural *land itself* for multifarious non-agricultural purposes. But the twentieth century is not the eighteenth or the nineteenth: the peasantry of the global South has nowhere to go to if it is dispossessed, in contrast to the dispossessed peasantry of the North that migrated in vast numbers to the New World.

The peasantry today is turning from passive forms of resistance like suicide to active contestation of the exercise of hegemony by global capital. This transition of segments of the peasantry and rural workers, from being passive objects to active subjects of history, marks an important moment of the current economic and political conjuncture. The present acute global food crisis is a direct outcome of the new phase of attacks on the peasantry, which has been going on for more than three decades but has escaped scholarly attention until very recently. In the following sections, we briefly discuss developments in India under neoliberal policies introduced from the early 1990s by way of sketching the background to the ongoing struggles against land acquisition. Starting with one or two cases a decade ago of opposition to state land acquisition for special economic zones, these struggles are now spreading rapidly.

Agrarian Distress, Farmer Suicides and Rising Unemployment

Two decades after neoliberal economic reforms started in India as part of the agenda of imperialist globalisation, the condition of the masses of the labouring poor on every objective indicator is worse today in every part of the country except where positive intervention has taken place to stabilise livelihoods (Delhi Science Forum, various years). On the other hand, the richest minority at the top of the income pyramid of this country are far richer than ever before and are better off than even the middle classes in advanced countries (Crédit Suisse, 2013), for they command extremely cheap services from the labouring poor, whose bargaining position as regards wages is lowered owing to rising unemployment; hence the constant addition to the reserve army of labour. The issues of greatest concern to those in the rural areas are: first, the attempt to take over their lands and resources by domestic and foreign corporate bodies, usually actively aided by governments; second, rising unemployment as high GDP growth fails to translate into jobs; and third, the high rate of inflation in prices of basic necessities, which is eroding their already low purchasing power (World Bank, 2014). The three phenomena are interconnected and they represent the results of pursuing the agenda of imperialist globalisation by implementing neoliberal policies.

The ruling classes in this country have long forgotten that 'development' means improving the well-being of ordinary people. They have long subscribed to the ideology of financial capital, which continues to play a hegemonic role despite global economic and financial crisis, and which entails an obsession with the rate of growth of GDP to the exclusion of any concern about how this growth takes place, how it is distributed and who it benefits. They seem to subscribe to a crude form of 'trickle-down' theory, in which if the rich get richer, the poor are automatically supposed to benefit through the increased demand for goods and services on the part of the rich.

What is actually happening is a dangerous combination of two trends. First, a drastic slowing down of the expansion of material production, especially in the vital primary sector, i.e. agriculture and allied activities, and particularly the key crop, food grains, which has seen falling per capita output (UNDP, various years). This has happened because, for the best part of two decades, through its policies of fiscal contraction and openness to trade, the state has actively attacked the small producers and created an agrarian crisis, which has by no means ended, but in some ways is intensifying into the struggle for land.

Second, the type of growth that has taken place has been acutely lopsided, with services now accounting for three-fifths of GDP, while agriculture and manufacturing have been relegated to contributing less than a fifth and less than a quarter, respectively (Reserve Bank of India, various years). Enrichment of the minority has meant a boom in construction and in eating out and travelling, including foreign travel, but construction and the hospitality sector are the only ones generating some employment, while in the main material productive sectors the job situation is dismal (World Bank, 2013). Minority enrichment has produced speculative real estate operations and an attack on the small property of farmers – in the name of development projects or special economic zones, which are but a front for real estate speculation, a pittance is paid for taking over farmers' lands, a process which farmers have at last started to actively resist (Alternative Survey Group, various years).

The share of agriculture, forestry and fishing in GDP has declined drastically, even though it still has to support two-thirds of the population, while the share of manufacturing has stagnated and that of the services has increased very fast. This is not a country that is 'developing' but one that is 'tertiarising' – it is becoming a modern version of the medieval economy in which dozens of service providers worked to maintain the high lifestyle of a tiny minority of the rich.

Third, because our farmers have been exposed to the extreme volatility of global prices, they have become indebted in the period of rising prices and have been unable to repay in the subsequent period of declining prices. Hounded by collecting agents of private moneylenders and banks, every year many thousands have been driven to suicide.

From the police records of suicides (which understate the actual numbers since not all suicides are reported), we see that farmer suicides have risen considerably more than non-farmer suicides (Nagaraj, 2008; see Table 5.1). By 2002, while the non-farmer suicides were 12.5 per cent higher than in 1997, the farmer suicides were 31.9 per cent higher. Another especially bad year was 2004, when farmer suicides were 33.9 per cent higher than in 1997, while non-farmer suicides were 16.1 per cent higher. By 2006, non-farmer suicides had also risen substantially. The global recession from 2008 and the drought in 2009–10 affected India badly, throwing many millions of people out of work, as the 2012 employment data show. The total number of farmer suicides starting in 1997 reached 250,000 by the end of 2011 (Ministry of Statistics and Programme Implementation, 2012).

Since land is often the collateral for loans, peasants have been losing land against unpaid debt during the period of agrarian distress under neoliberal reforms, and landownership concentration has increased, as shown by the National Sample Survey, which collects landholding data every ten years. At the all-India level, the top 5 per cent of households, ranked by owned area, accounted for 45 per cent of total owned area in 2003 compared to 38 per cent of total owned area held by the top 5 per cent in 1992. The share of every other group except the top 5 per cent has declined. The data for future years are likely to show a further rise in concentration.

Table 5.1   Number and Indices of Farmer, Non-Farmer and Total Suicides, 1997–2006

	Index of farmer suicides	*Index of non-farmer suicides*	*Index of total suicides*	*Index of total population*
1997	100.0	100.0	100.0	100.0
1998	117.6	107.9	109.2	102.0
1999	118.1	115.0	115.4	103.8
2000	121.9	111.9	113.3	105.7
2001	120.5	112.0	113.2	107.6
2002	131.9	112.5	115.2	109.5
2003	126.0	114.0	115.7	111.3
2004	133.9	116.1	118.6	113.1
2005	125.8	117.7	118.9	114.9
2006	125.2	122.9	123.3	116.7

Source: Nagaraj (2008). Available on www.macroscan.org

Note: Data on farmer suicides were not presented separately before 1997.

High GDP growth has not been producing jobs. This is bound to happen under capitalist production motivated by profitability and prepared to dispense completely with hiring labour if a machine can do the job and give them higher profits. The very fact of technological change and higher labour productivity means higher joblessness. Between 1993–94 and 1999–2000, the National Sample Survey data (see Table 5.2) showed that unemployment for men and women together had risen sharply in both rural and urban India as the state followed misguided policies of fiscal contraction and labour retrenchment under advice from the Bretton Woods Institutions. Some improvement then took place mainly owing to fiscal expansion during and after the bad drought of 2002–03: the economy saw 2.7 per cent annual growth rate of employment from 1999–2000 to 2004–05. The latest data for 2009–10, however, show that growth of employment has collapsed again more drastically than ever before, to only 0.1 per cent from 2004–05 to 2009–10, with a much sharper fall in rural areas (from 2.4 per cent to –0.3 per cent) than in urban areas (4.2 per cent to 1.4 per cent). Yet, the neoliberal policy makers, apparently blind and deaf to growing insecurity of livelihoods, continue to talk of carrying forward 'reforms', of further austerity, in order to placate international finance, regardless of the fact that these same financial interests have created recession in their home bases.

The construction of large dams, steel mills and manufacturing plants has been taking place in India since the late 1950s as a part of planned development, and the question arises why no sustained opposition to displacement took place at

Table 5.2   Annual Growth Rate of Employment 1987–88 to 2009–10

Period	1987–88 to 1993–94	1993–94 to 1999–2000	1999–2000 to 2004–05	2004–05 to 2009–10
Rural	2.6	0.8	2.4	–0.3
Urban	4.1	2.7	4.2	1.4

Source: 'Data on Usual Principal and Subsidiary Status (UPSS) Employment from National Sample Survey, 66th Round, 2009–10'. *Key Indicators of Employment and Unemployment in India,* in conjunction with Census 2011 data. Calculated by Chandrasekhar (2011).

that time, whereas now peasants and tribal people are strenuously resisting land acquisition and displacement throughout the length and breadth of the country.

One important part of the answer is that for nearly four decades after independence, up to the end of the 1980s, the state followed, on the whole, fiscally expansionary policies: it spent generously on investment and development, which kept the rate of growth of employment opportunities above the growth rate of the labour force. Those who were displaced through infrastructure or manufacturing projects did suffer distress, particularly tribal communities, but most were absorbed into new growing types of employment. This situation changed drastically as the Indian government embraced neoliberal policies from 1991, involving tight money, limits on government spending, reduction of the ratio of fiscal deficit to GDP, retrenchment of labour, devaluation, and removal of restrictions on foreign trade and investment. The immediate result, as explained earlier, was rising unemployment and loss of mass purchasing power, which has intensified in recent years with new rounds of domestic fiscal contraction and exposure to global recession.

In such a situation of disappearing job alternatives, the rural producer with a bit of land will naturally cling to it and will resist any attempt at dispossession. That bit of land is security against unemployment and destitution. No matter if the neoliberal attack on agriculture, combined with exposure to global price volatility, has caused acute agrarian distress and made farming so unviable – especially in the case of many export crops – that many thousands of farmers have been driven to suicide owing to indebtedness. Suicide is also protest, a passive form of protest. But this loss of viability of small producers was not because they did not work hard or because their land suddenly became less fertile. It was the result of totally misguided neoliberal policies, which reduced public investment drastically, withdrew extension services, made credit expensive or not available at all, and exposed small producers to the violent ups and downs of global prices by doing away with protection, while, at the same time, existing price-stabilisation systems were dismantled. Land as an asset did not thereby cease to be important – in fact in a scenario of jobless growth it became all the more important for peasants and workers.

Some Indicative Cases of Resistance to Land Acquisition

We may broadly classify the ongoing struggles of resistance to land acquisition or to change in land use into two types: large-scale land acquisition by the state governments for creating special economic zones (SEZs) to attract foreign capital; and relatively smaller-scale land acquisition for setting up extractive, industrial or infrastructure ventures by the state and the corporate sector, both domestic and foreign, the latter generally mediated through the state. With economic reforms there was a spate of SEZ creation, which did not initially meet much resistance. Landowners were intimidated by the existence of the Land Acquisition Act of the colonial period, amended several times, under which the government is empowered to acquire land for 'public purpose', and they were simultaneously lulled by promises of adequate compensation. As it became clear over time that, under the neoliberal dispensation, the so-called public purpose was to hand over their precious land to private corporations, which engaged more in real estate speculation than in setting up manufacturing, and as compensation did not materialise, the peasants started launching agitations, which has resulted in a number of them losing their lives in consequent police firing. The much publicised agitation during 2008–09 in West Bengal against land acquisition for setting up an automobile manufacturing plant, even though compensation was paid promptly, played a role in the loss of that state by the Left Front in the 2011 state assembly elections. In two states in India, namely Goa and West Bengal, in the face of opposition from farmers, the state governments withdrew proposals for setting up SEZs. In other states, however, farmers have increasingly come into confrontation with state governments on this question.

The central government passed the Special Economic Zones Act in June 2005. The number of operational SEZs by June 2012 was 143, while the number approved but yet to become operational was 633; thus giving a total of 776, with a relatively higher concentration in the South Indian states (other than Kerala). The total area under SEZ is not precisely known but is estimated to exceed 0.2 million hectares already (Alternative Survey Group, various years).

In a number of states which have a relatively high endowment of forest and mineral resources, and which consist predominantly of tribes, additionally the state governments have signed dozens of memoranda of understanding with foreign corporations permitting the extraction and shipment of minerals and precious metals. These projects involve substantial land acquisition and displacement, thus generating resistance.

In Odisha state (formerly called Orissa), the state government signed an agreement in 2008 with the Korean steel company POSCO for setting up a 12-million-tonne steel plant at Paradip, accepting the company's condition that it would export locally mined iron ore and import higher grade ore from Brazil. This agreement involved extensive takeover of peasant-cultivated lands in a large number of villages, land which produced valuable commercial crops and had given an adequate living to these farmers. The project has generated sustained resistance against land acquisition, which has brought it to a standstill. Even those farmers who had initially given up their land voluntarily have joined the resistance along with farm workers, since they lost their asset and principal means of livelihood years ago, while compensation has not been paid.

In Gujarat recently in August 2012, farmers from 60 villages in two districts campaigned against acquisition of their farmland for a proposed expressway and bullet train track linking the city of Vadodara in that state to the commercial hub of Mumbai. In Pune district of Maharashtra in August 2010, farmers agitating against diversion of irrigation water to a nearby industrial township were fired at and three farmers were killed. In Uttar Pradesh in May 2011, protests against land acquisition for roads in villages very near the Delhi national capital region led to clashes between local farmers and a force of nearly 10,000 police personnel sent to suppress them, in which four people died, two of whom were policemen.

There is a long history of agitation against raising the height of dams over rivers, since this involves submergence of more existing settlements, mainly in forested areas, and displacement of large numbers of peasants and tribal people. The failure to rehabilitate the displaced even after years have passed has led to a hardening of the resolve of the affected people to resist at all cost and demand time-bound rehabilitation in the form of land elsewhere. The most recent protest in Khandwa district of Madhya Pradesh state, in August–September 2012, has taken the form of *jal-satyagraha*, with protestors standing in neck-deep water for over two weeks. The concerned state government has agreed to the protestors' demands for reduction in the height of the dam and speedy rehabilitation of those already displaced.

Highly Variable Price of Land and Its Relevance for Peasant Anger

It is highly significant that farmers and rural communities are struggling against land acquisition and for compensation because it means that from passive forms of protest – suicide – they have turned at last to active forms of resistance. A decade ago this author, when drawing attention to the agrarian crisis long brewing in the countryside, was told that if things were actually that bad then peasants themselves would be protesting, which they were not. No one can put forward such an argument for ignoring agrarian distress now. Peasants are slow to move, but when they do start moving, no force can hold them back.

We have to understand that land has a special characteristic in that it is not a product of human labour. Though it is the cradle of all human activity, the extent of land cannot be increased beyond a point, once the limits of reclamation have been reached, while further deforestation would be highly detrimental. In this sense, land is a primary resource, which is fixed in supply. To paraphrase briefly what Karl Marx put in very striking terms in Volume 3 of *Capital* (Marx 1974: ch.46): since land is not a product of human labour, the 'price of land' is ultimately an irrational category. This is because of the different ways in which land prices and the price of commodities (including agricultural ones) arise. Commodities, which are all the product of labour, have prices anchored to values, namely, the amount of direct and indirect labour used for producing the commodity. Land price, however, is different, since land cannot be produced (though it can be improved).

What then determines the price of land? The price of land is based on the market capitalisation of the income the land will yield. And the fact is that this can vary enormously, depending on the specific use to which the land is put. This is an important reason for the discontent unleashed among farmers by land acquisition even when initially some may have acquiesced in it.

To understand this, let us consider a simple example. The price of 1 kg of rice of a given quality cannot possibly vary by a factor of 100, from Rs 20 to Rs 2,000, for example. But the price of a given hectare of land can vary from Rs 2,00,000 to Rs 20 million (namely, be 100 times higher, or even more) depending on the specific purpose for which it is used. If a farmer tills a hectare of land that produces crops from which he gets an annual net income of Rs 10,000, it is thought that he should find it reasonable to be offered a price of Rs 2,00,000 for that acre of land if the going market rate of interest is 5 per cent or more. For, if he then held that sale amount in a bank, he would get an annual interest income somewhat above or at least equal to the income he was getting earlier. This is the principle on which governments have been fixing compensation for land acquisition, namely, capitalising estimated annual agricultural income per unit area, at the going rate of interest.

However, once the farmer actually sells his acre of land to a developer, he finds that it is parcelled out in small lots to 100 urban buyers for residential or commercial purposes and the developer thereby makes an income of Rs 20 million, 100 times higher than the amount paid to the farmer. The farmer then feels thoroughly cheated out of a resource whose value, he feels, he did not know when induced to part with it for what now appears to be a mere pittance.

Since habits of investing money and living on interest are unknown to small-scale rural producers, and the state, though it takes away their asset with alacrity, does not bother to manage their cash compensation for them, the Rs 2,00,000 paid as compensation to the farmer becomes indeed a pittance, which soon disappears once a two-wheeler and a television set have been bought and living expenses met, leaving the farmer without any productive asset or source of income.

Our farmers have at last become alive to the trickery that is being perpetrated on them, of a land price determined apparently legitimately through the market, but actually taking advantage of their initial lack of knowledge. They are infuriated today at the idea of their precious land being forcibly taken away for a pittance in the name of development, especially when most of such 'development' does not create industry or provide jobs for them. SEZs and most smaller-scale land acquisition projects involve nothing but land speculation for commercial and residential buildings affordable only by the corporate sector or the rich minority of households. They see the 'developers' and the corporates make profits running into several hundred times the price at which they have sold the land. Even worse is the complicity of the state whenever it uses its power to acquire land to hand over to the corporate sector, often with a subsidy element, and takes no steps to ensure that the livelihoods of the displaced are safeguarded, beyond making vague promises. Worst of all is acquisition of good cropland to facilitate the setting up of ports for exporting mineral resources by foreign companies, as in the case of Korean POSCO in Odisha, for then two types of non-replaceable resources, minerals and cropland, are being looted.

The impact on agricultural production of indiscriminate cropland acquisition will clearly be adverse. By the early 1990s, the total cropped area in India had become stagnant, and in the following two decades, as public investment and development spending was cut back, the growth rate of crop production slowed down drastically, leading to falling per capita output of many crops, including the key crop, food grains (GRAIN, 2014). Further area diversion to non-agricultural uses will aggravate the supply problem. It is the longer-run neglect of output that already underlies the current food price inflation, which is high despite the substantial reduction in demand, which has been brought about at the same time through public spending being cut, and the impact of global recession, inducing higher unemployment and demand deflation.

On the other hand, it is also true that some forms of labour-intensive industry requiring land have to be promoted. There cannot be a permanent moratorium on land acquisition for non-agricultural purposes and nor do farmers themselves want all their children to engage in farming alone. The solution lies in creating land banks by identifying land not suitable for agricultural purposes and locating manufacturing enterprises on such lands. In cases – which should be rare – where some cropland acquisition by the government is unavoidable, farmers have to be treated as cooperative partners in the proposed venture by directly giving them a share in the assets of the enterprise concerned, over and above the cash compensation for the loss of land rights. Such arrangements are common in urban and peripheral areas of expanding cities, where householders occupying residential plots, who are literate and able to assert their rights, do not lose out from allowing developers to build multi-storey structures on their land, since in return they insist on getting not only cash compensation but ownership of up to one–third of the total built-up area, which appreciates in value over time.

In this era of the new primitive accumulation, the corporate sector, with the connivance of governments, thinks that it can, as in the past, continue to treat the small-scale rural producers as ignorant and exploitable dupes, taking their assets away and giving a mere pittance in cash in return. They are mistaken in this: the growing waves of resistance prove that assets can no longer be seized without providing the same or another form of asset in return. Finally, we need to remember anew Karl Marx's observation:

From the standpoint of a higher economic form of society, private ownership of the globe by single individuals will appear quite as absurd as private ownership of one man by another. Even a whole society, a nation, or even all simultaneously existing societies taken together, are not the owners of the globe. They are only its possessors, its usufructuaries, and, like *boni patres familias* they must hand it down to succeeding generations in an improved condition. (Marx, 1974: 776)

6

OCEANIA

The Papua Niugini *Paradox: Land Property Archaism and Modernity of Peasant Resistance?*

Rémy Herrera and Poeura Tetoe

Papua New Guinea (*Papua Niugini* in Creole), which gained independence from Australia in 1975, currently numbers about 6.5 million inhabitants. This population is spread over a territory of 463,000 sq km, which consists of two-thirds of the eastern part of the large New Guinea Island plus the surrounding archipelagos (New Britain, New Ireland, Manus, Milne Bay, Bougainville). For a long time, its social formations have provided an ideal ground for ethnologists and anthropologists, who found in its singular complexity an unending source of research, whereas few economists had turned their attention to it. Thus, the contrast is striking between the relative scarcity of economic analyses of this country and the numerous experts' reports, emanating from government agencies or more frequently foreign transnational corporations,[1] which are attracted by mineral resources (its gold reserves could place the country in third position on a worldwide scale, according to local estimates) (Department of Mining, various years),[2] hydrocarbon fields and even forestry resources, and are massively investing in the region.

The land question is a real issue in this peasant society – characterised by a population consisting predominantly of farmers and breeders, having a domestic economy largely based on subsistence, with the persistence of 'customary' systems and autonomous local territories integrated into the space of a 'nation state'. Indeed, it is a source of recurrent and sometimes violent conflicts with regard to landownership and use among transnational corporations, the state and the civil society – the latter has often been demeaned by archaic prejudices.[3] Thus, it is one matter to stress the existence of a landownership that is not founded on private proprietorship of soils, and of collective forms of land cultivation and production whereby a range of rights are placed side by side, applying to individuals and run by various institutions and practices emanating from the state, modern authorities and 'customary' communities; it is another matter to consider all these factors as insurmountable impediments to 'development' in a capitalist sense. Conversely, if the 'undevelopment' of this Southern country is undoubted, such a phenomenon cannot only be explained by external openness, which is adopted under pressure and uncontrolled.

The original feature of Papua New Guinea is that community lands still cover more than 90 per cent of the country (Table 6.1) – somewhat incredible in a dependent and neoliberalised economy (Corrin Care and Paterson, 2008 and Lalau, 1991). The peasant resistance has a lot to do with it (Herrera and Tetoe, 2012a). In this chapter we try to analyse how internal factors (connection between Papua New Guinean society and the state) and external ones (connection between the state and foreign capital) are linked to prevent, in fine, the development of the country. After a brief demographic and cultural presentation (first part), we study the traditional land institutions and the relationships to land – frequently considered to be 'archaic'– (second part), then the steps of the process of registration of land during the colonial period and since independence (third part), in order finally to examine the surprising modernity of this peasant society and its capacity for resistance (Cooper, 1994; Scott, 1985, 1990) to the unusual forms of neoliberal policies[4] (fourth part).

People and Culture

Although a set of questions remains to be addressed about the exact origins of the Melanesian people (see Pawley et al., 2005), archaeological and linguistic discoveries in the last few decades have shown that the settlement of Papua New Guinea originates from successive arrivals from South-East Asia (Gille and Toulellan, 1999; Hope and Haberle, 2005; O'Connor and Chappell, 2003). More than 40,000 years ago, the first wave brought in the hunter–gatherer ancestors of the Papuan population, who settled in various parts of the northern coast of New Guinea, then crossed over to the neighbouring islands of New Britain and New Ireland, and then to Bougainville (by 30,000 BP) and Manus (by 20,000 BP). These populations had earlier settled in the lowlands, and then progressively penetrated into the rugged mountainous valleys of the Central Highlands, where better climatic conditions permitted land cultivation. The practice of irrigation in agriculture, for yam, sagu, taro and sugar cane, for instance, has been shown to have existed for at least 9,000 years (i.e. since the end of the last glacial period), making Papua New Guinea one of the centres of early agriculture (Bourke, 2009; Denham et al. 2003; Denham, Haberle and Lentfer, 2004; see also Golson, 1991; Sullivan, Hughes and Golson, 1987).

Furthermore, the development of an intensive agriculture led to a concentration of populations in the Highlands. A second large migration occurred in 10,000 years BP, bringing Austronesian people to the coastal lowlands of the great island and the surrounding archipelagos. All these migrations have led to an exceptional diversity in languages;[5] socially and culturally speaking, this makes any generalisation hazardous.[6] However, many observers have suggested the existence of commonalities. Papuan groups share the same attachment to ancestor worship mythology and beliefs as well as forms of artistic expression that characterise the entire island of New Guinea, and more generally the Melanesian area. The lack of hierarchy within the socio-economic organisation of population groups has often been taken as demonstrating an autonomy and adaptability in traditional institutions that puts the authority of the state into perspective.

The scattering of these communities does not mean that they were separated from each other or, as is often assumed, only came in contact to fight. In the Highlands, most of them maintained exchange networks that connected them with remote groups, as revealed by the salt trade, for example. Ceremonial exchanges, such as *Moka* and *Tee Cycle*, were opportunities for hundreds of groups – and individuals – to compete with each other for prestige and influence through their *big men* (Godelier, 1982). The difficulty of assimilating a clan or a group into a 'policy', the relative but real autonomy of individual behaviour among those communities, and their linguistic differentiation are elements that might support the notion of a social formation on the verge of collapse and engulfed in persistent internal conflicts; this in turn has led to the view that a national sentiment cannot emerge in Papua New Guinea.[7]

The European colonisation, whose main actor was Australia – this 'England from the Antipodes', as Fernand Braudel (1981) wrote –, was slow in coming[8] but has had a great influence on populations that are historically and culturally diversified. The colonial legacy is heavy, notably because of the artificial partition of the country. The New Guinean island is cut into two parts. Its eastern part is the independent state of Papua New Guinea, which was formerly known as British Papua – then as Australian Papua from the 1906 transfer on – in the south-east and German New Guinea in the north-east. The two were reunified in 1949 following the PNG Act, thanks to which Australia took full control of the Eastern part of the island, including the former German quarter as well as the territories temporarily occupied by Japan during the Second World War. The western part, Papua (ex-Irian Jaya), a former Dutch colony, has been part of Indonesia since 1962. Separatist tensions still shake some Melanesian islands, as in Bougainville, where conflict degenerated into a tragic war from 1989 to 1998 (Watts, 2007). The racial distinction drawn between 'black islands' (Melanesia) and islands of Polynesia had far-reaching effects on the collective European imagination and fantasy: the violent 'black cannibal' in the West;[9] the apolitical and isolated 'noble savage' in the East.

This European penetration slowly and partially integrated the indigenous people into global capitalism, transforming most of them into small farmers and keeping them dependent on colonial plantation companies, particularly for the production of palm oil (in Sepik and New Britain) and coffee (in the Highlands). Yet, almost three-quarters of the Papua New Guinean workforce is still involved today in growing subsistence crops. A considerable proportion of what is produced is for self-consumption or sold in local markets, and only a very small amount is exported. One of the most significant distinctive features characterising this peasant society is its connection to land and persistent attachment to traditional institutions to defend collective landownership – even if it is not possible to oppose collective ownership to private ownership in a simplistic manner.

Connection to Land, Customary Rights and Collective Ownership

In Papua New Guinea, as in most societies of the South and particularly in Oceania, land is only exceptionally subject to private appropriation, as modern capitalist right means it (see Tables 6.1 and 6.2). This customary management, which excludes private property but secures for all its members access to collective land, does not by any means result in egalitarian access, since differentiations exist among families and clan leaders and between the genders. Customary practices are not free from social hierarchies and prejudices. Thus, we tackle the question of customary rights in this peasant society neither with naivety and nostalgia nor with undue panegyric.

Under customary law, some members within clans and families are designated as leaders to administer lands. Generally, family ties give access

Table 6.1 Distribution of Land by Kind of Property in the Oceanian Area (percentage)

	Public property	Private property	Customary property
Federal States of Micronesia	35	1	64
Fiji	4	8	88
Kiribati	50	5	45
Marshall Islands	1	0	99
Nauru	10	0	90
Papua New Guinea	2.5	0.5	97
Samoa	15	4	81
Solomon Islands	8	5	87
Tonga	100	0	0
Tuvalu	5	0	95
Vanuatu	2	0	98

Source: Corrin Care and Paterson (2008).

Table 6.2   Proportion of Registered Customary Lands in Oceania
(percentage of total lands)

Country	Proportion of registered customary lands
Fiji	+90
Federal States of Micronesia	+50
Kiribati	+50
Marshall Islands	−10
Nauru	+90
Papua New Guinea	−5
Samoa	+20
Solomon Islands	−1
Tonga	0
Tuvalu	100
Vanuatu	n/a

Source: Corrin Care and Paterson (2008).

Notes: +x = more than x per cent; −y = less than y per cent; n/a = no available data.

to land, but very often the only effective way of organising collective work (clearing, upkeep of gardens, etc.) was to let someone exploit land temporarily, without however allowing them to sell or rent it (see Herrera and Tetoe, 2011). Furthermore, this interest in land does not confer rights to its exclusive use and occupation, as other communities may also share rights on it. Hence, the result is a singular and complex economic system in which very personalised relations of production still remain today, providing a vision of individual labour lying within protection or association pacts and turning towards self-consumption and basic-needs satisfaction; trade dominated by exchange of prestige and barter; an apprehension of money separated from market prices, hoarding and accumulation of capital (Godelier, 1982).

In Papua New Guinea, land law draws from two main sources that are admitted by the Constitution: on the one hand, there is the Western capitalist tenure system of 'alienated' lands (freehold and leasehold), which was introduced by the colonial administration and is currently followed by the modern state (see World Bank, 1978); and on the other hand, there is the customary land tenure. Currently, alienated lands, and especially those used for mining exploitation, represent only 120,000 parcels, or 600,000 hectares: less than 5 per cent of the national territory.[10] More than 90 per cent of Papua New Guinean land corresponds to customary land. According to the 2000 census, about 80 per cent of the total population is still living in rural areas. Nevertheless, land is always the object of desire of private interests and under pressure for registration and privatisation. More and more parcels have been registered as customary lands by the Department of Lands.[11]

Facing pressures from foreign investors and international donors, the position of the state of Papua New Guinea seems rather ambivalent. Theoretically and legally, all customary land transactions are controlled by the state. The persistence of customary land use rights, which is an issue for foreign capital, causes serious problems for public authorities in granting mineral rights to transnational firms or allowing the building of infrastructures to boost private foreign investment. These require both concessions of resource exploitation rights to private investors and the alienation of customary land (such alienation could be unlimited and exclusive) by the state (Anderson, 2006). However, we can observe that the dominance of traditional collective forms of social organisation within the peculiar structure of land tenure in this country has not prevented, but only slowed down, the growth in exports of minerals, hydrocarbons and agribusiness products.

It is true that some land transactions are beyond state control. But, it has been noticed that when the state acknowledges landownership of families and clan leaders – even when the soil is rich – its protective role over customary land use is effective only if private interests are not involved and no natural resources have been discovered. If this is the case, the state takes over the land to sell the exploitation of all resources existing on or below its surface: minerals (gold, copper, nickel, etc.), oilfields and gas fields, forests and even water. Access to natural resources and their exploitation by foreign transnationals are thus carried out with the support of the state of Papua New Guinea, which articulates this process of land appropriation with the previous ancestral structures of collective landownership without introducing 'free' land markets.

When a resource project is launched, compensation solutions generally involve financial payment to landholders by mining and petroleum companies as well as royalty payments by the state. In addition, local communities might also have direct equity involvement in all major resource developments. If the law does not require the preliminary identification of all beneficiaries, the distribution of compensations and royalties is legally dependent on the registration of landholders. The jurisprudence on land affairs has de facto encouraged competition between landholders, leading some of them to adopt a logic of 'ideology of landownership' (see Filer, 2006; Filer, Burton and Banks, 2008) and rent-seeking strategies (Banks and Ballard, 1997; Filer and Imbun, 2009; see also Strathern, 2009). Even in Papua New Guinea, where traditional social structures remain strong, the majority of the peasant society has been receptive and reactive to such financial incentives (Fingleton, 2007).

However, the monetisation that followed has not necessarily led to the collapse of the social structures as the effects on them were complex and contradictory. Obviously, self-interested behaviours have spread, leading to strained relationships and even disintegration of families or clans. But simultaneously, groups that were identified as landholders – or even as equity holders – have earned not only money but also prestige and honour. Maurice Godelier has already demonstrated – in the 1980s, when about two-thirds of the peasant families were involved in the monetary economy – the singular way in which the Papua New Guineans spend money to maintain or extend social relationships (Godelier, 1982), particularly on weddings. From an economic point of view, money is not usually saved or invested but is spent on the social competition of gift giving. Social structures have not collapsed but have rather adapted. Over the last few decades, the distribution of royalties by transnational firms to the central government, local governments and landowners has favoured the landowners.[12] Even if more and more land disputes are brought before the courts, most complaints are related to the amounts of compensation paid by private investors rather than to landownership.

In contrast, a revendication from the peasant communities is frequently the jointly held ownership, as a measure for preventing dissolution of groups (division between brothers, among others, for example). In fact, despite constant and convergent pressures from foreign transnationals, the governments of developed countries and international institutions towards an individualisation of landownership, the successive Papua New Guinean authorities have not been able (nor have they wanted) to challenge the customary collective landownership. The main reason is popular resistance mounted by the peasant society against the privatisation of lands, the imposition of a modern register for lands and their management by the capitalist laws. Consequently, if the customary land system has not represented a conclusive obstacle to the neoliberal strategy applied by the government towards the opening up to foreign capital and the growing exploitation of natural resources, the people of Papua New Guinea are nevertheless vested in legitimate landownership.

Collective Ownership under Threat of Land Laws: Some Historical Facts

The registration of customary lands and the establishment of cadastral systems had always been the first priority for the Australian colonial administration (Crocombe, 1987; Dale, 1976; James, 1985; Williamson, 1997). In inland areas that had been recently 'discovered', pressures exerted by the colonial authorities on local groups had brought about frequent confrontations between the colonists and the natives (Antheaume et al., 1995; Neale, 2005). The colonial plantations and trading companies, from Southern Australia (or Germany) arrived in force on the coastal plains and surrounding islands, where alleged 'vacant' lands (the expression used by the colonial power for lands that were uncultivated and unoccupied) were grabbed by private companies, especially for the cultivation of copra. The progressive opening up of the Highlands offered opportunities to the Australian administration to conquer more areas. Parcels were requisitioned and rented out to colonists, who created a workforce from among the local people, in particular for coffee cultivation. Coffee cultivation increased, and along with it monetisation, among communities including those in the Highlands. In a limited number of cases, access to land use was managed by customary authorities under colonial state supervision (employing traditional leaders) in order to oblige peasants to produce the pre-established quota of export cash crops – a system also employed by the Dutch in Indonesia.

Most of the time, the aim was land grabbing and, more generally, the direct control of land by the colonial authorities. After the Second World War, mining explorations and exploitations led to the most important penetration into the territory (Herrera and Tetoe, 2011). From the end of the 1950s, Australian mining companies started arriving in hordes. This investment inflow into mining sped up in the 1960s, supported by Canberra and the World Bank, and then turned towards the petroleum sector. Following the opening in 1972 of the Panguna Mine by Bougainville Copper – the Australian subsidiary of the London-based giant conglomerate Rio Tinto – gold and copper from 1976 onwards became Papua New Guinea's largest export items, well ahead of copra, coffee and cocoa. This expansion further consolidated the country's productive and commercial specialisation in primary activities and increased its reliance on Australia, whose domination on these sectors was crushing. In return, today, Papua New Guinea is the largest recipient of Australian aid in the region, enabling the country to equilibrate its balance of payments and to cover its public deficits. The highest priority of the state has always been, and remains today, to attract foreign direct investment in strategic sectors like mining (gold and copper), hydrocarbons (petroleum) and also agricultural cash crops (coffee, cocoa, copra, palm oil, etc.). In order to achieve this, the institutional environment has been 'secured' by reducing the risk of expropriation and by minimising taxations (Bank of Papua New Guinea, various years; Economist Intelligence Unit, various years; see also Herrera and Tetoe, 2012b).

In 1952, the Native Land Registration Act established the Native Land Commission, which was given responsibility for customary land registrations and dispute settlements. Ten years later, in 1962, this institution was replaced by the Land Titles Commission, which pursued the same goals but with more modest ambitions. At that time, strict control over land transactions was implemented. The Land Act provided that 'a native has no power to sell, lease or dispose of customary lands otherwise than to natives in accordance with custom' (Section 73). It was left to the government to purchase or lease customary land (Section 15), and a principle of unanimity had to be applied according to which all members of a landowning group had be involved in approving any transaction. In the event, the results of these systematic land registrations were poor, forcing the administration to think up new incentives. The new idea was to encourage communities towards voluntary transactions.

By virtue of the Land Tenure Conversion Act (1963), if all members of a group of landowners agreed, one member of the group could individually apply for registration of a freehold title to the land. However, fewer than 1,000 conversion orders were made under this Act in 1965.

Faced with the landowners' resistance, the Australian administration finally came to the decision to secure their rights on acquired lands. A package of Land Titles Bills was adopted from 1969 onwards, supplemented by additional Land Reform Bills in 1971. This new offensive launched by the Land Titles Commission against collective landownership planned to 'clarify' and move towards simplifying the customary tenure, while strengthening controls on land transactions. As the export specialisation of Papua New Guinea was evolving from agriculture to mining, the laws aimed to establish the most appropriate land tenure to gradually articulate the country into Australian production and trade structures. The bills were intended to operate geographically selective registration of interests in customary land. Group interests were to be registered in the names of representatives, who were given the power – but not necessarily the right – to deal with lands as if they were their absolute owners (Trebilcock, 1983). But finally, these bills were withdrawn in the face of widespread objections, and the so-called 'reform' was suspended. Some of this popular opposition focused on the risks of tenure individualisation and land aggregation in an elite's hands, with landlessness and landlordism as likely consequences (Bruce, 1988).

In 1972, Papua New Guinea became self-governing. Following the withdrawal of the land laws, the new authorities set up the Commission of Inquiry into Land Matters, which reported in 1973. Its report recommended a range of reforms related to numerous issues of land policy, including the reduction of inequalities in the allocation of landownership, control of private freeholds, the prohibition of selling land to a foreigner in areas characterised by land shortages, and the setting up of a decentralised system of dispute resolution at the district level. Leasing to non-citizens was limited to a 60-year term. In 1974, the government also declared its intention to acquire land alienated by the Australian colonial administration and colonists by implementing the Plantation Redistribution Scheme, under which plantations owned by expatriates or foreign companies had to be nationalised (except for those with a strong capitalistic structure, like tea or palm oil plantations, or cattle breeding farms). Furthermore, communities were offered the opportunity to acquire some plantations with the financial support of the state: the group taking over the land was expected to pay a deposit of 10 per cent of the plantation's value and to transfer to the government the remainder within between three and five years. At the time of its independence in 1975, Papua New Guinea embedded in its Constitution the recognition of local groups' landownership.

Yet, 'national' development was experienced for such a short duration that the country seemed to move straight from colonialism to neoliberalism (Herrera and Tetoe, 2012b). The Papua New Guinean economy still depended deeply on the world system, especially on Australia, the main regional power (Amarshi, Good and Mortimer, 1979).[13] The dynamics of trade and financial exchange remained unequal and polarising, strengthening the influence of a capital ownership dominated by foreign transnational firms. These dynamics were entirely dictated from the outside and did not meet the country's needs of consumption and development. The only possible way for the state to implement an economic policy favourable to offering concessions that allowed foreign investors to exploit natural resources, and to increasing exports of cash crops, was the commitment to an advanced process of customary land alienation (Weiner and Glaskin, 2007) in conjunction with the registration process initiated in colonial times.

Under the Lease–Leaseback Scheme of 1979, the customary owners could nominally lease land to the state, which provided titles that allowed the custodian to lease the land to a third party. Such a mechanism has boosted the development of transnational agribusiness through the settlement of smallholders. In 1981, the Land Registration Act continued the registration of land and legalised new private landowners (forbidding confiscation by banks). This trend expanded during the 1980s, leading to the Land Administration Improvement Programme in 1984 and the Land Evaluation and Demarcation Project in 1987, with the support of generous financial aid from the World Bank, the Australian Agency for International Development (AusAID) and the United States. At the same time, pilot land registration experiments were attempted in several areas – for example, the East Sepik Land Act in the Sepik region (Fingleton, 1991; Kavanamur, 1997; Power, 1991).

All these developments were the start of a conclusive offensive carried out by SAPs. The neoliberal policy of opening up to foreign direct investment, which was adopted by the Papua New Guinean government, had been curbed by the extent of lands that are exempt from capitalist laws and depend on ancestral systems of communal property. Many observers and experts have considered this peculiarity as the decisive factor among restrictions affecting the private sector, with examples listed in the World Bank reports (see, among others, World Bank, 1978, 1989).

In a context of external imbalances (balance of payments, indebtedness) and under the aegis of the IMF and the World Bank, the government of prime minister Sir Rabbie L. Namaliu undertook the first SAP in 1989. As elsewhere in the South, the 'reforms' implemented consisted of privatising national companies, cutting public spending, freezing wages and making the labour market more flexible, restricting the monetary policy and devaluing the currency (kina), dismantling tariff barriers, liberalising external trade and making the economic national territory more 'attractive'. In Papua New Guinea, however, the SAP's keystone was landownership: the implementation of a modern legislative frame for registration and division of lands. As a prerequisite to loans and grants awarded by international institutions, the Land Mobilisation Programme ran for five years from 1989. With the support of Australia through the Australian Contribution to the Land Mobilisation Project, this programme was extended in 1993 for an additional three years (Iatau and Williamson, 1997).

In 1995, a second SAP, called the 'Structural Reform Plan' – a name adopted to give the impression that it had been possible to dismiss some of the IMF's diktats – was imposed by the government of Sir Julius Chan. The reality was a deepening of the country's economic submission. At the core of the measures that were taken – which included the adoption of a floating exchange rate, privatisations, weakening public services (health, education) and the lowering of wages – the reactivation and speeding-up of the land registration process was listed again under the designation of 'land reform' (i.e. land privatisation). Later, in June of the same year, the main sections of the Customary Land and Registration Bill were disclosed. As a sine qua non for international grants and loans, the plan provided a legal framework for finalising and implementing land registration in the provinces of East Sepik and East New Britain (Larmour, 2003).

A surge of protest began across the country, forcing the Ministry of Lands to stop the programme and constraining the IMF and the World Bank to accept the withdrawal of the land issue from the SAP. To calm people down, a Land Group Incorporation Act was introduced. The calm was quite brief since, as early as 1996, the Land Mobilisation Programme was being extended. Moreover, a series of Acts was promulgated by sector, aiming at easier access to lands and exploitation of natural resources. The Mining Act (1992) and the Oil and Gas Act (1998) are particularly important as the rights of landowners to use their lands were laid down and finally limited (article 113 of Oil and Gas Act). The third SAP was launched in 1999, following the external repercussions of the Asian financial crisis (1997–98), and put back at the core of 'reforms' the land issue, again with the financial backing of the IMF, the World Bank, Papua New Guinea's close neighbour Australia, and its ally the United States (via the USAID),[14] joined by the Asian Development Bank and Japan. In 2001, the government of Sir Mekere Morauta gathered a group of experts to draft a bill on the privatisation of customary land. But again, popular demonstrations ruined the bill.

Social Issues, Peasant Resistances and Popular Mobilisations

The social indicators of Papua New Guinea – which has the third highest gold reserves in the world – are amongst the lowest in the world. According to UNDP estimates, 38 per cent of children were provided with access to schooling at the beginning of the 2000s (UNDP, various years; World Bank, various years). The literacy rate was 63.9 per cent against 47 per cent 30 years earlier. Even with the linguistic diversity that characterises the country making the task of the education authorities more difficult, it is clear that public expenditure on education remains inadequate and that the neoliberal offensive launched against public services has caused a withdrawal of the public educational sector in favour of the private one (notably religious and international). With regard to health too, the indicators are very poor. The infant mortality rate was close to 80 per 1,000 live births in the year 2000, whereas life expectancy at birth was still below 60 years. There were barely seven doctors per 1,000 inhabitants. Vaccination rates remained lower than average in the Oceanian subregion.

Concerning food consumption, in spite of nutritional deficiencies that at times are serious, cases of chronic and severe malnutrition are quite rare in Papua New Guinea. One explanation might be that the large majority of the agrarian population has access to community lands and there exists a practical support system in operation, based on the redistribution of income from collectively grown subsistence crops (named the *wantok* system), helping to reduce inequalities, cushioning the effects of the structural crisis and, in some cases, preventing social decay. These traditional support mechanisms, however, are less in evidence and less effective in urban areas, where such social ties tend to become weaker or are likely to be rebuilt using other methods. Consequently, poverty has increased, especially in towns, where almost 70 per cent of the population was said to be living below the poverty line in the mid 2000s. Being poorly developed, the public health care service is declining compared to private structures, which are supported by the state and international organisations but are out of reach for the poorest Papua New Guineans.

Following from this, one can easily foresee what devastating impact the expansion of capitalist landownership can have on the survival of these ancestral and institutionalised solidarity ties, and the reasons why some sectors of society are rebelling. As a matter of fact, the recent popular uprisings in this country have not only involved peasants *sensu lato*, but also workers' organisations from the mining industry, civil servants (notably teachers' unions) and student movements, women's associations, and even some (impoverished) soldiers of the army (see Imbun, 2000; Herrera and Tetoe, 2011). And for some years now, more and more researchers have chosen – rightly so, according to us – to study the complex and fast-evolving contemporary Papua New Guinean society in terms of 'class' (for instance, in economic anthropology, Gewertz and Errington, 1999).

As stated earlier, the neoliberal strategy, which was implemented by dominant capital on a worldwide scale, has stumbled in Papua New Guinea (without however being halted)[15] over the absence of a unified ownership system based on a generalized form of private land property; the recommendations of successive SAPs for the registration and division of collective lands have always caused considerable popular discontent. Before the neoliberal era, massive resistance had taken place against the penetration of transnational mining companies. The most virulent action was that exerted by people from the island of Bougainville against the conglomerate Conzinc Rio Tinto of Australia, forcing the closure of the gigantic gold and copper mine of Panguna in May 1989 and worsening a latent and recurrent secessionist conflict, which later turned into an open war between the Papua New Guinean armed forces – loudly supported by Australia – and the Bougainville Revolutionary Army (BRA). Despite its singularity, in the sense that it was amplified by a separatist movement, and because of the violence of an unequalled aggression against local populations in the country, the Bougainville conflict created a precedent that was followed by numerous community groups anxious to assert their fundamental rights before the state and transnational corporations.

Resistance to successive governmental 'reform' projects of the status of customary lands, imposed by foreign capital interests and with the support of local elites – including the military – has lain at the very core of the most popular mobilisations organised during the last few years. In fact, demonstrations by communities have not ceased to increase, expand, organise and intensify over the greater part of the country. Protesting people fear that the possible division of collective lands would lead to weakened control of clan leaders over their territories and loss of traditional values, along with the deepening of social inequalities and the emergence of new landless peasants.

One example among many is the case of groups from New Britain fiercely rejecting the World Bank's and the Asian Development Bank's programmes aimed at increasing cultivated areas of high-yield palm oil plantations.[16] Production of palm oil, of which Papua New Guinea is now one of the world's leading producers (after Malaysia, Indonesia, Nigeria, Thailand and Colombia), is devoted to exportation and to supplying transnational corporations involved in the agro-food sector and synthesis chemistry, especially in the refining of biofuel.[17] Another example arose in the Enga Province in April–May 2007. Following demonstrations by peasant communities, the mine at Porgera, one of the largest gold and copper deposits in the world, had to shut down; (one should also bear in mind that the Panguna mine has not reopened since Rio Tinto, the main shareholder of Bougainville Copper, was forced to leave in 1989). Since 1999, a violent conflict also broke out close to the Indonesian border at Ok Tedi, in the Star Mountains of the Western Province, between Ok Tedi Mining, majority-owned by BHP Billiton (the largest mining company in the world and the first one listed by market capitalisation in Australia), and the local communities scandalised by the devastating effects of open-pit mining on the environment (Herrera and Tetoe, 2011).

Faced with such popular uprisings, the Papua New Guinean state adopted an ambivalent position. On the one hand, it alternated between demagogic 'good intention' statements and the negation (against all evidence) of any will to reform the customary land tenure; and, on the other hand, it reasserted the attractiveness of the national territory for foreign capital, fiercely negotiated royalties or compensations payments, and resorted to the repression of protesters. A lot of experts and intellectuals believe that landowners are worrying over nothing, since Papua New Guinea has recognition of customary land titles embedded in its Constitution; many others think that land registration is essential for development and that traditional ownership obstructs it (Curtin and David, 2006; Iatau and Williamson, 1997).

As for international institutions and donors, they openly encourage all practices that are in favour of a new land tenure system – such as the Land Mobilisation Programme, and other more specific projects, that have been adopted since the World Bank's recommendations in the 1960s. It seems to be difficult for some people to understand that this resistance is much more than people merely harping on about an antiquated past. This resistance expresses the defence of inalienable rights of access to land and its collective use for the well-being of communities systematically attacked by neoliberalism for three decades. Besides, it conveys a legitimate revolt against ecological crimes caused by intensive natural resources exploitation by transnational companies.

In addition, a crucial point to understand is that these peasant resistances are articulated alongside other more global claims from many components of Papua New Guinean society – workers of the private sector, civil servants, poor urban people, students, environmental activists, feminists, even religious progressives – confronted with the SAP austerity and with privatisation of the common patrimony. The climax of anti-neoliberal protests, gathering together diversified but convergent social movements, was reached with a series of demonstrations in 2000–01. One of these movements came from the army, and notably from the regular troops based in Port Moresby. They contested a decision taken by Commonwealth experts – directly deriving from the SAP framework – to reduce military budgets and strength. The army protesters demanded the eviction of all Australian military advisors present on national territory, foreign mercenaries under contract with the Papua New Guinean government or employed by transnational corporations to secure strategic sites (such as mines and oilfields), as well as IMF and World Bank experts. Hundreds of people joined these demonstrations in the streets of the capital. A few months later, in June 2001, a walk of several days was planned by student associations, again in Port Moresby. They protested against the neoliberal policies of the government, chanting 'Rausim IMF, World Bank, Australia!' ('Out IMF, World Bank, Australia!'). When the procession of demonstrators, including thousands of trade unionists, clan leaders and activists from progressive associations, headed towards the

Parliament, which was secured by the army, some soldiers spontaneously joined the movement.[18] The police repression that followed forced the government to suspend privatisation programmes of public enterprises (among them, post and telecommunications, public transport and banks) and to give up its customary land 'reform'.

Conclusion

The time has come to reconsider the idea and perception the rest of the world may have about the popular struggles in Papua New Guinea, which appear more modern, and in many respects more radical, than those elsewhere. Western mainstream media often interpret the struggles in the wrong way, considering them to be no more than expressions of irrational fear in the face of supposedly ineluctable socio-economic change: the opening up of domestic markets to foreign investors, privatisation of public enterprises, and above all the breaking of the system of customary landownership. According to the economic mainstream, private alienability of landownership at market prices is considered to be the optimal way of managing this means of production, indeed the only rational way of using it, at both individual and social levels. Consequently, in practice, the methods Western modernity used for the expansion of capitalism worldwide – that is, the deprivation of the peasantry – should be universalised to include the Southern people.

In a peripheral economy like that of Papua New Guinea, the complex system of landownership, combining the rights of the state and the persistence of customary practices, might have had the effect of serving the interests of a savage capitalism, leading the country into dependence and increasing inequalities. The challenge is to resist the violence of neoliberalism that has been imposed for decades, as well as the exploitation of natural resources. What is being defended is the legitimacy of the principle of collective landowning and free access to peasant community land; what is being demonstrated is the possibility of other rules for land use; and what is being suggested is to maintain the existence of a non-capitalist peasant farming.

These struggles were given less coverage in the media than the views on insecurity expressed by the dominant classes – and a fraction of the middle classes that have profited from neoliberalism. Various prejudices obliterate the reality of the dependence on Australia and the struggles of a people longing to master its collective destiny. In our analysis of the economy of Papua New Guinea, we have stressed the potent constraints on it. In a country where the gross domestic product ($15 billion) hardly exceeds half of the annual turnover of one of the numerous transnational companies working in its ground (as in the case of Rio Tinto with a turnover of more than $25 billion per annum), the government has little room for manoeuvre. But options exist. An alternative to neoliberalism is sought, along with the emergence of a 'class alliance' around peasantry, to achieve a modern development strategy that benefits the Papua New Guinean people.

Notes

1. Many of these experts' reports are written by distinguished academics – some of whom are listed in the References – who are acting as observers of public authorities such as the Australian Agency Aid or as consultants for private mineral interests. For an informative analysis, see Stewart and Strathern (2005).
2. For an analysis of foreign capital and the mining exploitation issue in Papua New Guinea, see Herrera and Tetoe (2011).
3. Unfortunately, several examples are found in the daily press, especially from Australia, and academic reviews.
4. We define 'neoliberalism' by referring to the running of the capitalist world system and by giving it a class meaning, as 'the doctrinal system on which is developing the global strategy of domination of high finance, together with the institutional and ideological superstructure under its control'. On this point see Herrera (2010a).
5. Almost 750 languages and three vernacular languages (Tok Pisin, Hiri Motu and English) have been listed (Schroeder, 1992).
6. There are more than 1,000 different cultural groups in Papua New Guinea. See Foley (1986) and Pawley et al. (2005).
7. According to Defert (1996), the word 'Papua' apparently came from the Molucca Islands and could mean 'fatherless', highlighting the absence of centralised power.
8. For the most isolated regions, it began around 1930, or even later.
9. These prejudices were revived by certain tragic events, such as the death in 1961, during an expedition in Asmat in the south-west, of Michael C. Rockefeller, a young heir of the wealthy US dynasty.
10. At the beginning of the 1990s, of the total land area of 47,615,700 hectares, Lalau (1991) found 435,100 hectares to be private freehold land (1 per cent), 870,200 hectares, state land (2 per cent) and 46,310,400 hectares, customary land (that is, 97 per cent).
11. See the information disseminated by the Department of Lands, and section 6 of the Mining Act (1992).
12. For many people, however, compensation is not a solution. Some groups have asked for the removal of national money and the return to the *kina*, that is, the ancient exchange of traditional shells.
13. On the mechanisms of the dependence on Australia, see Herrera and Tetoe (2012c).
14. The US Agency for International Development is the 'cooperation' agency of the federal government of the United States of America.
15. Herrera (2006). For a Marxist analysis of the current crisis, see also Herrera (2011, 2012, 2013a, b).
16. Since 1971, the World Bank has encouraged the production of palm oil in New Britain (Hoskins), and, since 1976, in East Sepik, allegedly with 'the support of local landowners'. See Bourke (2001, 2005); Bourke and Vlassak (2004).

17. A press release by local communities of New Britain said: 'We, the landowners, are developing and will continue to develop *our land* on our own terms. We therefore sternly warn all those parties involved in wanting to use our land for oil palm to *stay out*! Any attempt to bring oil palm on our land will be strongly resisted.'
18. See, among many others, the *Sydney Morning Herald* (June 2001).

7

EUROPE

An Overview of the European Peasants and Their Struggles

Gérard Choplin et al.[1]

At the heart of a continent that is rich overall compared to many other parts of the world, the majority of European farmers live in difficult circumstances (in 2007, the number of farm workers remaining in the 27 European Union member states was 26,669,000, working in 13,700,000 farms). Most have incomes lower than the minimum wage of other professional categories. The repeated sectoral crises have eliminated the weakest farms, most often those of small or medium size.[2]

In the so-called developed countries in the 'North', the peasants now represent only a small percentage of the active workforce (in 2008, 5.4 per cent in the European Union as a whole; 1.4 per cent in the United Kingdom; and 1.7 per cent in the United States). Is this a sign of 'development'? In a Europe in which unemployment is soaring due to the financial and economic crisis, are we moving towards a world without peasants?

Europe has a rich diversity of agriculture, rural crops, agricultural products and regional or local foodstuffs. It is also very diverse in its agricultural structures, its modes of transmission of farming from one generation to another and in the history of its peasant struggles. The recent inclusion in the European Union of ten countries from Central and Eastern Europe that were for 45 years in the Soviet zone has reinforced its diversity and complexity.

As in other continents, European peasants have for the last 25 years been victims of the results of the neoliberal wave that has globalised agriculture. With agricultural prices often below the costs of production, the value of the peasants' work and the costs of agricultural products are not recognised, which demotivates young people from becoming farmers. European agricultural subsidies intended to compensate these low prices primarily benefit a minority of large farms and agribusinesses and perpetuate dumping into the Third World countries. 'You can't touch "free trade". It's cancer, and it's deadly', says Christian Boisgontier, former spokesman for the Farmers' Confederation in France (Via Campesina, 2010).

Today, the symptoms of the neoliberal disease and of overproduction appear in the news: Earth is getting warmer, hunger is increasing, biodiversity is disappearing, fossil fuels are running out, unemployment is soaring in Europe, and the gap between the rich and the poor has never been greater in most countries in the world. The cost of maintenance of this excess with injustice will become exorbitant, political spaces will open up to all those who resist, and the European farmers are part of it.

We are probably at 'a historic junction, where the worst and the best are possible, where the unexpected can occur, the repressive logic as well as creative potential.'[3] Later in this chapter, we see how European peasants and citizens are acting to escape from this society that is sick from its addiction to overproduction.

Struggles are taking place in Europe on agricultural inputs (land, seeds, etc.), modes of production (high productivity, integration, the environment, GMOs, health, climate, etc.), agricultural policy, and the rules of international trade, among others. These are linked to similar struggles in other continents – 'globalise the struggle, globalise hope', says La Via Campesina, the worldwide peasant movement.

Indeed, this is not to pit the 'North' against the 'South'. To put it in a nutshell, the confrontation is rather between two outlooks of agriculture and commerce: the outlook of the WTO on the one hand, and that of food sovereignty on the other. Farmers in the North and the South, though their conditions of life and production may be very different, are confronted with the same takeover by transnational agribusiness and trade of agriculture, food and agricultural and trade policies. The globalisation of agriculture is a social, environmental and societal failure.

What are the most important issues for European farmers today? Are farmers going to disappear, becoming mere subcontractors without any rights within the food industry? Will they remain confined to certain markets or become engines of a relocated economy? The European Union is once again an importer of food. Will it continue to outsource an increasing portion of its agricultural production to 'low cost' countries? European farmers, diminished in the West by neoliberal capitalism and destroyed in the East by Soviet communism, are aiming for productivity. Will they rediscover or develop anew their multifunctional nurturing role in a reunited Europe, as is needed by European society?

The 'left' has always had ideological difficulties in taking the peasantry into consideration. It has often legitimised the family farm but has also announced its death as a result of the inevitable advance of technology. Having repeatedly accompanied gone along with the increasing submission of agricultural production to market law, will it finally give high priority to the peasant question?

> Since the revolution in social relations is seen as intrinsically linked to the industrial revolution and development of the working class, the peasantry appears as a leftover of history that can only survive if it gets rid of its small-scale producer ways. But this self-employed worker, master of most of his means of production, who lives from the fruit of his work and who does not exploit any outside labour but of his family, represents fundamentally anti-capitalist values. He is seen as such because he appears to be a symbol of contradiction in the midst of this worldwide system. The inability of different socialist schools to define its place in the society they want to build leads them in fact to defend the current form of agricultural production. The difficulty of matching the economic theory and the analysis of social relationships explains the ambiguity of the socialist approach with respect to the land question. The implementation of the revolutionary project thus seeks to recreate a peasant specificity that the doctrine rejects. This contradiction explains the relative weakness of the socialist parties in the campaigns and why they have really not been able to politically mobilise farmers. (Duby and Wallon, 1976: i. 11–12)

In 2012, the European Union decided to reform its agricultural policy for the period after 2013. Will it learn from the past decades? Are there alternatives to the current agricultural and trade policy that could give a future to farmers in Europe and elsewhere?

European Farmers Struggling with Industrialisation and the Globalisation of Production and Consumption

Farmers are vanishing

After the Second World War, when Europe experienced hunger, and with the creation of the European Economic Community, a European agricultural policy was mooted, aimed primarily at ensuring food security in Europe and at upgrading, strengthening and restructuring agriculture to bring the workforce to the expanding sectors of industry and services.

The increase in productivity per farmer, per hectare, per animal, was one of the strongest in all economic sectors. Farmers were asked to produce more despite being fewer in number. Subsidies were paid to them to cease their operations so that those of others could be expanded. And many farms could not survive the agricultural prices, which, though guaranteed, were fixed in terms of high productivity that the farmers failed to achieve. After 1992, severe price cuts and the vastly unequal disbursement of direct aid continued to push the concentration of production into a number of increasingly smaller holdings. The large number of old farmers who left the job was not balanced by younger people coming in; the young moved out because of lack of acknowledgement, both economic (agricultural prices were less than the costs of production) and social (social rights were far less than those of other professional categories). In half a century, agricultural policies pushed out four out of five farmers.

The rapid disappearance of farmers in the North, who were the majority of the population less than a century ago, is undoubtedly one of the important events of the twentieth century. It ended an era that started in the Neolithic period when agriculture was invented. Is it really the end? The trend could be reversed if there is political will amongst governments, the European Parliament and the European Commission. This shift in outlook would also have to extend to the economic actors and regulatory, legislative and financial sectors, bearing in mind their negative role in the maintenance and development of peasant agriculture.

Agribusinesses destroy everyday peasant agriculture

Productivism has developed farms that are increasingly dependent on firms or cooperatives tied to the upstream, downstream and banks. In breeding, economists found that there were no economies of scale beyond a certain size: for example, in Denmark, a country of intensive and large-sized farms, the cost of a litre of milk or of a pig is the highest in Europe. Another example is provided by the agricultural situation of Brittany, which, paradoxically, despite the concentration of half of France's livestock in 6.5 per cent of its national territory, is ranked second to last amongst the French regions in terms of added value.

The interests of agribusiness are usually far removed from the expectations of farmers and of society in general (in terms of climate, product quality, jobs, etc.). After having 'selected' farmers, the 'logic' of business is now trying to take society hostage: to defend the ongoing restructuring of agriculture in the name of international competitiveness, this is to defend agriculture in a world that suffers from hunger and malnutrition. It does so with the support of producer groups, which, instead of being an extension of the farms to enhance the product, turn into logistical support for the concentration of production. The recent mobilisation of leaders of the major French cooperative groups (such as Cooperl and Coopagri) denouncing social dumping by Germany, where labour without minimum wages is cheap, well illustrates the situation we are in.

The development of agriculture and the livestock industry in Europe and worldwide is leading to a rapid decline in peasant agriculture. The reform of the Common Agricultural Policy (CAP) currently under discussion should not be a simple adaptation of the current policy, but must break away from the increased funding of large farms by agribusiness,[4] which provokes a degree of competition that has disastrous effects within the farming system. In the current context of CAP reform, the redefinition of agricultural activity in relation to industrial activity in Europe is very important. This reform should legitimise support for small-scale agriculture and redefine the roles of producer organisations.

Non-reproducible industrialised farms

In countries like France, where each generation has to buy again the farm and all its equipment, farms are too costly to be taken over by youngsters (especially for livestock farming, which requires heavy investment in buildings

---

Box 7.1   **Who Benefits from the EU Agricultural Subsidies?**

Before the 1992–94 WTO agreement, European agricultural subsidies were used mainly to buy surplus production generated by the productivism mentioned earlier, storing it and exporting it to the world markets. At that time, European agricultural prices were higher than world market prices and the European Union had to pay subsidies to exporters to sell their surpluses. Subsidies benefited mainly storage and export companies, as well as large producers, who could overproduce without any limit at guaranteed prices.

After the WTO agreement, the EU lowered its guaranteed agricultural prices to the global market level, which depended more or less on production. The goal was not to escape from paying export subsidies but to avoid being accused of dumping. However, as European production costs were higher than world agricultural prices, the EU provided direct aid to producers to compensate for the fall in agricultural prices. This aid, having no ceiling per farm, increased the EU budget and quickly became a significant part of the income of farmers. Indeed, as the EU has kept exports as one of its priorities, it continues to export agricultural products at prices below the cost of production. Thus, the dumping continues as always – formerly with export subsidies and now with direct aid.

Does this aid benefit farmers? First, it partially compensates for world prices, which are generally kept artificially low (due to dumping). Second, it is mainly used by agro-industry producers to purchase agricultural products at prices lower than the cost of European production, which, it goes without saying, amounts to protecting the European market and financing the imports of agro-industry and the supermarkets, which do not pass lower prices on to the consumers.

This is a dirty trick played by the European Union and the United States and codified in the Marrakesh Agreement: it allows them to continue the dumping in a new form vis-à-vis Third World countries, while protecting themselves from imports. Third World countries are not foolish; the EU has ruined the legitimacy of the CAP at the international level and the Doha Round is dead. The latter is a good development, but without questioning the rigged Marrakesh Agreement, there will be no way out from the effects of the crisis for European farmers and those from Third World countries.

and machinery). These farms are frequently in debt and become victims of the recurrent sectoral crises and the banks, which can stop their credit.

The milk crisis of 2009 has led many small, medium and large dairy farms to bankruptcy and too many farmers to suicide. Fifteen percent of the milk production in the UK has been abandoned. In its place, the farming firm Parkham Farms tried to install a 'milk factory' of 8,100 dairy cows in Lincolnshire; this project was stopped by citizens and the authorities.

Outsourcing of European agricultural production

One goal of globalisation in agriculture and the offensive against customs duties is to allow agribusinesses to supply from anywhere in the world, or to grow for themselves wherever production costs are lowest and then transport the products to countries whose purchasing power allows them to sell at a high price.

This is how flower production has left Europe in a big way for Colombia, Ecuador, Kenya and India. While there are thousands of vineyards in the European territory, large European investors are planting vineyards in South Africa or Chile. Production of organic fruits and vegetables, which require more labour, has also been relocated to Turkey, North Africa, and other countries. The poultry industry, a major consumer of soybeans, after having devastating environmental and social effects in some parts of Europe, has now shifted to Brazil. More lamb now comes from New Zealand than from Europe, while more and more beef will come from South America if the so-called 'free' trade agreement with Mercosur were to be signed.

When some say that access to European markets for farmers in the South is a condition of 'development', they forget that it is primarily European companies that are relocating in the South. The Southern countries need the access to the European markets to be without tariffs so that they can 'repatriate' their products. And, the first market to which farmers in the South would like access before they consider exporting is their own local, regional market.[5]

It is thus responsible for the cultivation of 16 million hectares of soybeans (equivalent to the combined agricultural area of Germany and the Czech Republic) in large monoculture plantations in South America, the majority from GM seeds, which has social and environmental consequences just as devastating as those previously experienced in Europe. Intensive farming in Europe is hence very dependent and very fragile, even though Europe could grow its own vegetable proteins – particularly legumes, which would also save a lot of nitrogen fertiliser.[6] European agriculture is standing on its head. In Romania, in the village of Mosna, 90 per cent of dairy cows have disappeared since the country joined the EU in 2005, and organic farms of 30,000 hectares aimed at the Western European market are flourishing (ÖBV, 210).

What happens to European farmers in such a context?

EU governments have gradually become aware of the lack of international legitimacy of their social and environmental agricultural policies. But due to the lack of political will to tackle the problems at the root, they have, since 1992, provided environmental aid and animal welfare for rural development, etc., to try to mitigate the damage from the deregulation of markets and to legitimise their policy in the eyes of their own people and the Third World countries. In 1999, in Seattle, they used the 'multifunctionality' of European agriculture as a selling argument, but without success, and now they have had to agree to the payment of 'public goods'.

From what has been stated earlier, European farmers occupy a great variety of situations and statuses. We find, among others:

- poultry or pork breeders, or vegetable growers, fully integrated into the agribusiness, producing like contract labourers, but often with fewer rights than factory employees;
- cattle and sheep breeders, for whom public assistance represents more than 100 per cent of their unassisted income;
- industrialised breeders, highly attached to the agro-industrial complex but very fragile in their relationships with the banks;
- large farmers (cereals, oilseeds, sugar, etc.), taking unfair advantage of all direct aid, expanding at the expense of smaller farms and creating a social desert around themselves (through rural depopulation);
- medium farmers, still numerous but often condemned either to grow at the expense of their neighbours or to disappear, especially when nobody wants to continue farm operations (the average age of farmers in Europe is very high);
- pluri-active farmers who, due to the necessity of survival, because they are too small, or prices are too low, or as a life choice, take up a parallel profession, or transform their products so they can be sold directly to consumers, or rent out accommodation, and so on;
- farmers specialising in regional quality products and known through their farm labels, which can attract better prices for their products.

We are in a multi-speed Europe:

- Seashores near the ports of soybean import, where livestock production is concentrated, become populated, while other regions are deserted.
- Some areas are polluted by high levels of intensive agriculture or animal husbandry, while others are transformed into nature parks.

- In Central and Eastern Europe, small subsistence farms exist right next to former collective farms of thousands of acres with dozens of employees.

The new European commissioner for agriculture, Dacian Ciolos, said that all these have their place in agriculture in the future CAP; but experience shows that some forms of agriculture are destroying others. We need to tackle the machine of productivism and 'free' exchange if we want to keep the farmers in Europe.

European Farmers in Struggle

The inertia of agricultural organisations connected to economic power

The majority of farmers belong to national agricultural organisations that are members of the Committee of Professional Agricultural Organisations (COPA)[7] at the European level. These organisations are most often associated with cooperatives, agricultural banks and agribusiness firms, and frequently have presidents who wear several hats and are active both in the labour movement and in the economic sphere. They usually belong to or are associated with political parties, often the Christian Democrats, who have governed many European countries since the Second World War. The organisations are generally corporatist, focusing primarily on defending the interests of their members and their economic allies and forging weak partnerships with civil society. In some countries there is genuine co-management of agricultural policy by the government and the 'classic' agricultural organisation, which frequently claims the monopoly of representation. In many countries, agricultural services such as insurance, social security and applications for European grants are supported by these agricultural organisations, thus forcing the farmers to join them. So we should hardly expect these organisations to challenge the 'established order'. Usually, they oppose any reform of the CAP, putting pressure on governments to ensure that the reform does not affect the privileges of large farms, but then defend it once it has been adopted.

Emergence of European and international farmers' movements

Despite the agricultural policy that was in effect and the standard organisations that were defending it, farmers managed to rise against them. In the 1970s in France and Germany, young farmers became aware of the social damage being caused by productivism and tried to influence the stance of agricultural organisations, though without success. They then created farmers' organisations in opposition to the official organisations. In the early 1980s the major problems facing the CAP (surplus, cost) led to small farmers' organisations from several countries meeting regularly and coordinating their efforts.

Thus, in 1986, the European Farmers Coordination (CPE) was born. This has expanded to many countries to become the only European farmers' organisation, in opposition to COPA, to present alternatives to the neoliberal policies pursued by the EU. From the outset the CPE decided not to be corporatist and has forged links with civil society, since agricultural, food and rural issues affect the whole of society.

When the location of important decision making on European agricultural policy moved from Brussels to the international level (WTO) during the Uruguay Round of negotiations, the need was felt for peasants to have an international voice. Thus in May 1993 in Mons (Belgium) the CPE, together with farmers' organisations from other continents with whom it had established contact over the previous ten years, launched an international peasant movement: La Via Campesina.[8] By 2008, the CPE had grown to become the European Coordination Via Campesina (ECVC) (see www.eurovia.org).

From local and national struggles of farmers to European and global struggles

Many farmers' struggles have developed in Europe, some of which will be reviewed later in the chapter. It is worth noting that because of the huge diversity of situations and histories in Europe, these have different characters in different countries. For example, in France it is relatively easy to occupy without violence the property of a big business that wants to expand, because public opinion will be a priori more favourable; whereas the same action in Germany would primarily be seen as an affront to private property.

Farmer's struggles at the European level are more difficult to conduct than those in individual countries. The wide range of situations among countries whose priorities are rarely the same and the lack of European media to publicise these struggles hinder their actions. Furthermore, the significance of European intervention has decreased since the CAP was taken over by the WTO. Today, it is easier to be the citizen of a region or country, or of the whole planet, than of a Europe that has lost its economic borders with globalisation.

In Central and Eastern Europe, the problem is that there are still no farmers' organisations. Forty-five years of Soviet communism and forced collectivisation of agriculture have immunised today's farmers against the idea of a 'collective'. But for how long? Agricultural organisations have emerged, often created by political parties or by business. Farmers' struggles in the future will probably arise from civil society, with associations opposed to big industrial farms, or to GMOs, or in defence of local seeds, as has happened in Romania. At the global level, European farmers have played a significant role in La Via Campesina in the struggles against the WTO for new global food governance (within the framework of the Food and Agriculture Organization of the United Nations, for example), with mobilisations against G8/G20, in the World Social Forum, at conferences on climate and biodiversity, etc.

Struggles for land

While in many African and indigenous communities in the world, land is a common good, for a long time this has not been the case in Europe, apart from communal grazing, which is progressively disappearing. Individual private property developed in Europe, especially during the nineteenth century, after the French Revolution. The modes of transfer and inheritance of land and the laws for the leasing of farmland differ from one European country to another. In some countries, the mode of inheritance has fragmented the landscape into tiny plots. In others, the rules privilege the continuation of large farms.

In France, egalitarian inheritance has exploded the number of landowners, creating a market for the sale and lease of land and rendering those who worked on the land insecure, as the property right is absolute and cannot be challenged. The law concerning rent (lease) after the Second World War gave new rights to farmers vis-à-vis the owner, which means they no longer have to buy land and go into debt to secure their businesses.

Numerous local struggles against the accumulation of land by large farms have marked recent decades (occupation of land, symbolic installation of a young farmer cultivating a plot, etc.). These have sometimes succeeded in helping small farming operations to start up or expand.

The struggle of the farmers from Larzac in France provides an innovative response to the question of the right to work the land (Via Campesina, 2010). From 1971 to 1981, 103 farmers protested non-violently against the French army, which wanted to extend a military camp onto agricultural land. They won after a massive mobilisation throughout the country. The state had already expropriated 6,300 hectares, which were later freed. The Société Civile des Terres du Larzac (SCTL) was created and signed a 60-year lease with the state, renewable on expiry (see SCTL, 2002). This contract offered farmers a 'career lease' on the land at the usual price of tenancy, which was not automatically transferable to the descendants. It is the SCTL that collectively decides on the allocation of land or buildings that have become available.

Another recent initiative in France is the organisation Terre des Liens,[9] which has implemented a new form of access to land acquisition by making small areas throughout France collectively available to organic producers. Here too, use takes priority over ownership. These projects are being developed primarily in peri-urban areas and supply directly to consumers.

In Europe, there are huge gaps in farmland prices. In the Netherlands and Belgium population density and urban pressure make agricultural land expensive, usually more than ten times more expensive than in France. In Galicia (Spain), although the land is generally poor, the price is high because each family is attached to its land and there is little available for sale. With such large differences in price, particularly between countries in Western Europe and Eastern Europe, farmers can sell small plots in the West to buy large farms in the East, which has been 'colonised' in recent years. In Denmark, because of the huge industrial livestock production, land is especially sought to spread manure, leading to concentration of land into larger farms. The organisation Frie-Bonder Levende Land acts to oppose this process of concentration. In some poor areas near urban centres, entire villages have been abandoned and the land has reverted to forest and fallow. Is this temporary? If Europe relocates its agricultural economy and stops producing cheaply in the South what can be produced at home, these areas may well acquire new life.

The struggles of rural and urban youth to become established and to promote local food

Given the aging of the farming population in Europe, the difficulty faced by young people in gaining access to the means of production, and a general lack of social and economic understanding of working in the fields, few young people want to become farmers. Recently, however, Western Europe has seen the emergence of youth movements, often loosely organised, which seek to maintain peasant farms and increase the number of young farmers. This campaign is supported by La Via Campesina at the European and international levels.

These groups are part of farmers' organisations or rural youth organisations like the International Movement of Catholic Agricultural and Rural Youth (MIJARC). They want to take over their family's farm or are rural minded and looking to settle as farmers or food artisans. Others are more urban: they demonstrate the interest of the urban youth in the agricultural world and, especially, in a mode of food production linked to a different life philosophy. There is also a new interest in agricultural production and sustainable action in the city: for example, in the development of community gardens. These movements are growing in strength and are attracting hundreds of people in Europe from all walks of life and backgrounds. One of them, called Reclaim the Fields, organised several symbolic actions in 2009 and held a camp of 300 young Europeans in France. Together these movements are fighting for and exploring new forms of peasant farms: individually or collectively owned, involving new types of marketing, etc. These youngsters are fighting for a radical overhaul of the system. They are indeed aware that to establish a large number of young farmers on farms of a human scale, operating modes of production that are environmentally friendly, substantive changes will be needed in several fields. One not only has to address issues related to access to the means of production, but also to change the relationship between producer and consumer, allowing greater intermingling of the fabric of rural and urban areas. In other words, the whole system of trade and trade relations is brought into question here.

In Central Europe, which inherits a very different historical situation, little interest has so far been shown in these movements.

The fight against GMOs

Industrial rationalisation applied to agriculture has led to major malfunctions: specialisation, monoculture and failure to rotate crops have favoured the emergence of invasive weeds, insect pests, etc. But the headlong rush continues to generate more and more hybrid seeds, chemical fertilisers, pesticides and genetic (transgenic) technologies. Genetically modified plants (GMOs) produce insecticides or are tolerant to herbicides and are presented by agro-seed companies (Monsanto, Syngenta, Bayer and others) as the only solution to feed the world. But hunger is increasing, not because of problems of food availability, but due to lack of access to food (speculation, unequal distribution of wealth) and to problems with agricultural production. People become dependent and lose their food autonomy as the power of agro-industry increases.

Today, the struggle to maintain agricultural farming requires an outright rejection of the use of genetically modified plants and animals. As evidenced by the unhappy experiences of organic farmers in Catalonia, the culture of GM corn contaminates neighbouring fields. Contamination is inevitable and irreversible, and there is no possible coexistence between GM agriculture and agriculture free of GMOs. Patents related to these GM seeds impose the control of GMO firms over seeds and destroy what remains of peasant autonomy.

Fortunately, all over Europe farmer organisations and citizens are uniting and resisting GMOs. Despite frantic lobbying by the agro-industry, 80 per cent of Europeans continue to reject GMOs. For 15 years, widespread campaigning, acts of circumvention leading to numerous lawsuits, and hunger strikes have helped in preventing the cultivation of GM crops in Europe. Until 2010, only one GM maize had been authorised, but very little of it has been cultivated anywhere (with the exception of a few countries, such as Spain). In France, a moratorium has halted the cultivation of GM plants.

Many European regions that are united under the 'GMO-Free Regions in the EU' association refuse GMOs on their territory. However, the battle is not over. Every day tonnes of GM soya come into Europe to feed livestock. If it is not in the fields, it is still on our plates.

If we want to win a Europe free of GMOs, we must obtain:

- a complete ban on the cultivation and importation of GM foods in Europe;

- the prohibition of any industrial property right on living things, whether seeds, animals or genes, as well as of patents and the 1991 Plant Variety Certificate that aid industry in transforming farm seeds into counterfeit seeds;
- a common agricultural policy for achieving self-production of European vegetable protein (to stop the importation of soybeans).

Fighting for seeds

Peasants are the ones who select all the plants we eat. The seed industry has tapped into this immense diversity of plants to develop channels adapted to petrochemicals and have thus replaced 90 per cent of the plant selection made by peasants with a wasteful and polluting fossil fuel. By restricting the seed market to only the varieties homogenised and stabilised by the industry, competition from farmers' seeds has been eliminated; these seeds die away and disappear from the fields, surviving only as collections of 'plant genetic resources'.

The industry also wants to prohibit farmers from re-sowing a portion of their harvest. Many battles have been conducted since the 1980s to protect that right and the right to sort the crop to use as seed. In 1994, a European regulation recognised this right for the 21 species grown, but immediately transformed it, with the Plant Variety Certificate (or *Certificat d'Obtention Végétale*), into an 'exemption' of property right conferred on the industry, where the exemption is subject to the payment of royalties.

Since then, the industry has faced resistance from farmers. To recover royalties, each breeder must prove that a peasant has planted its variety and has not done her/his own reseeding. In Germany, the ABL union won in court the peasants' right not to respond to the injunctions of the industry, which wanted to force them to answer a questionnaire indicating the names of the varieties planted. In France, an inter-professional agreement allows the collection of royalties on deliveries of soft wheat harvest, but the peasant mobilisation has restrained parliamentarians from extending these royalties to other species.

Today, more and more farmers want to get out of the vicious circle of 'more debt to buy more chemical and mechanical inputs to produce more and sell for less'. Citizens concerned about healthy food and environmental protection encourage them. But without fertilisers and pesticides, industrial seeds do not produce the promised crops. Farmers then try to replant some of the crops that are fit for local production conditions and climatic variation. But most industrial varieties are either F1 hybrids that are not reproducible or are too specialised to evolve towards more natural conditions of culture. Farmers then return to 'old varieties' still cultivated or enclosed in collections. After a few years, these varieties of crops that have never been genetically manipulated begin to give interesting results with few inputs, even in less fertile land and despite years of climatic stress, whereas the industrial varieties wither. This work can only be collective. Establishing local seed houses often involves working with gardeners and civic associations; such networks are coordinated in each country and at the European level.

The industry has so far not admitted defeat. With GMOs, it has been attempting to introduce patents in seeds. Its aim is clear: a simple laboratory analysis enables it to recover royalties by proving that the peasant is reproducing its patented gene, including when it has not been purchased from any seed industry and the farmers' variety has been genetically contaminated. To circumvent opposition to GMOs in Europe, the industry is filing patents on genes from other unregulated genetic engineering. It is also preparing widespread 'genetic identification filing' (or *fichage génétique*) of plants. More immediately, it has been trying to persuade legislators to impose an additional condition for the access of farmers to the market, or to CAP: that of the identification of varieties planted.

Fifteen years of fierce farmer and citizen resistance have slowed the progress of GMOs, which have not been able to invade Europe. The new battles against industrial property rights and genetic filing of seeds, to establish a GMO-free Europe and to give farmers the rights to sow, exchange, sell and protect their crops against biopiracy and genetic contamination, will determine the right to food sovereignty.

Struggles against neoliberal agricultural policies and the current rules of international trade

European agricultural policy change affects not only European countries but also many others. From 1986, first the CPE and then the ECVC worked with European Union institutions and the public through meetings, hearings, conferences, the campaigns of their member organisations in different countries, and the European Days of Action.

On 26 November 2007, a demonstration took place in Brussels before the council of ministers to request a thorough reform of the CAP. In March 2009, with many other national and European civil society organisations, the ECVC launched a 'European movement for food sovereignty and another Common Agricultural Policy' to promote CAP 2013 and strengthen the momentum around food sovereignty in Europe. To this end, the Nyéléni Europe Forum was held in Austria in August 2011.

Without changing the current rules of international trade, it will be difficult to make changes to the CAP that move things in the right direction. That's why Europeans need to struggle at the global level against the WTO, 'free' trade, transnational or genetic food, and for new global food governance. The European Coordination Via Campesina is striving to reach these goals with its global movement. Mobilisations against a new WTO round, from Seattle to Cancun, Hong Kong and Geneva, have been successful in delaying this new cycle of negotiations, which lies buried today due to the upsurge of global crises that have undermined free trade.

Another victory belongs to all those who have fought the Economic Partnership Agreements (EPA). The EU wanted to force the ACP (Africa, Caribbean, Pacific) countries to sign trade agreements that were very unbalanced. ACP farmers' organisations, governments, farmers' organisations such as the European ECVC, and many development NGOs have worked to defeat the EPA.

To remedy the failure of the WTO, the European Union has signed or is negotiating 'free' trade agreements with several dozens of countries around the world, and it is difficult for civil society to mobilise against so many agreements. But today 'free' trade, which is anything but free, is buffeted by the global crises that it has caused. The European Union–Mercosur negotiations, for example, which could condemn most European cattle businesses to bankruptcy, might well be unsuccessful due to opposition from many governments and organisations.

Perspectives and Alternatives

Global crises open up a space for alternatives

The neoliberal model is now reaping what it has sowed. Major global crises are affecting the planet and its people, and they cannot be resolved without changing the direction and priorities of the model.

Giving priority to transportation, import/export, faster consumption of disposable products and over-exploitation of natural resources generates symptoms that affect the entire globe. A window of political space has opened up to promote alternatives that are challenging both socially and environmentally. With regard to European farmers, these alternatives include:

- making food sovereignty a part of agricultural policies (see Box 7.2);
- promoting farming as a purveyor of employment, a well-nourished population and respect for the environment;
- campaigning for global food governance;
- participating in international mobilisations on climate, biodiversity and against 'free' trade agreements, the WTO, etc.

A citizens' alliance for food sovereignty

In Europe, the public wants to maintain a predominantly peasant and multi-functional agriculture. To change the agricultural, food and commercial policies that concern all citizens, farmers, now reduced to a minority in society, have participated for the last 20 years in creating local, national and European alliances with organisations working in the fields of the environment, consumption, economic development, animal welfare, democracy, human rights, etc. It remains to strengthen them and to make them converge. Thus, a 'European food declaration' was signed in 2010 by over 335 organisations from 27 European countries.[10]

Local initiatives for food relocalisation

Throughout Europe initiatives between producers and consumers are multiplying to relocate food production. Abandoned, vacant lots of urban gardens are being occupied by small groups of citizens to grow vegetables. Associations are set up in which consumers pay producers in advance to produce what they want. Consumer groups subscribe to weekly baskets of fruits and vegetables from local producers; farms sell directly to the public at farmers' markets. Many municipalities facilitate these processes. Schools and company canteens seek to obtain local organic food. While these initiatives are very small compared to supermarkets sales, they are growing, and consumer demand is often greater than what local producers are able to supply. Environmental issues, product quality and health concerns all encourage consumers to seek out local sources of food.

There is still much work to be done and strategic discussions to open in order to implement food sovereignty, both at European and international levels, and changes will be needed in both policy and law. Even though the food sovereignty movement has gained much ground in 14 years, neoliberal opponents know how to denigrate it, distort it, and qualify it as 'protectionism'. But it has become, both in the North and the South, an effective force for reunification and social mobilisation.

'Another common Agricultural and Food Policy in Europe is possible', states the ECVC and its allies. This requires abandoning the current rules of international agricultural trade and the current uses to which the EU agricultural budget is put, replacing the priority given to import/export with that of feeding the European population.

This idea revolves around three objectives:

- to maintain and develop sustainable and social peasant agriculture that feeds the people, preserves the environment, promotes health and keeps rural landscapes alive (for this, farmers must be able to live primarily off the sale of their products, through stable and remunerative agricultural

Box 7.2   The position of the ECVC on Food Sovereignty – January 2010

Food sovereignty gives people and the EU the right to define their agricultural and food policy based on peoples' needs and their environment rather than according to the rules of international trade as laid down by 'free' trade ideology. For example, it is up to the EU to outlaw growing or importing of GMOs if the EU citizens do not want them, without the WTO having any say in the matter. Food sovereignty sets the priority for agriculture to feed people first and foremost, rather than producing for international trade. The EU has become the greatest importer and leading exporter of food produce, and therefore needs to totally reconsider its priorities. Exporting milk powder while simultaneously importing soy to feed cows, growing fruit and vegetables – even if they are organic – in the countries of the South because labour costs are lower there, all lead to the current social and environmental failures. Food sovereignty, on the other hand, relocalises agricultural production close to where consumers live. Food sovereignty, by allowing farmers to play a central role in feeding people in their region, provides them with a sense of social legitimacy that has often been lost through the current CAP. Food sovereignty is opposed to the current concentration of 'food power' that lies in the hands of agribusiness and supermarket chains. It is the duty of political powers such as the EU, for example, to regulate production, markets, and distribution, and to take all the actors in the food chain into consideration. It is also up to producers and consumers, as is increasingly the case, to shorten the chain through a variety of forms of direct sales. They should be encouraged to do this by the agricultural and food policy (CAFP) and safety standards for products processed on the farm – now industrial standards – should be adapted.

But make no mistake: food sovereignty does not mean autarky or a retreat behind borders. Nor is it opposed to international trade: all regions of the world have their own specific produce that they can trade; but food security is far too important to allow it to depend on importation. In all regions of the world, the basic food should be produced locally where possible. All regions should therefore have the right to protect themselves against low-cost imports that destroy their home production.

Food sovereignty not only confers rights, it also implies a duty to not damage agricultural or food economy in other regions of the world. All forms of dumping, i.e. all grants that allow exporting products at a lower price than the production cost should be forbidden.

Food sovereignty is aimed not only at feeding today's population but also feeding future generations, and therefore at the preservation of natural resources and the environment. This is why we need to develop modes of production which decrease agricultural emissions of greenhouse gases and benefit biodiversity and health. By cutting down on transport and shifting away from over-intensive agriculture, we are dealing with the environmental and climate challenges. Food sovereignty can provide a meeting point for all those in Europe who are working to change agricultural and food policies and those who are working for the relocalisation of food. This is the dynamic that can add weight to the orientations of the future agricultural policy.

See 'Pour une Politique agricole et alimentaire commune dans le cadre de la souveraineté alimentaire' (www.eurovia.org/spip.php?article273).

prices, rather than having to depend on subsidies; this is a necessary condition for their economic recognition, and will lead to greater attractiveness of the profession in the eyes of young people);

- to reserve public support for sustainable production and farms, which are beneficial for employment and the environment;
- to relocate food production as much as possible and stop the takeover by retail and food-chain industries.

The world requires European policy to change. On every continent, agricultural policies must be freed through food sovereignty and new rules of international trade, with regulation replacing speculation. We cannot meet the immense food, social and environmental challenges unless our cultural patterns of 'modernity' change. One should no longer need to say that 'a modern society is an urban society with few farmers', but rather that 'a modern society is a society with more farmers and food artisans in a relocated economy'.

Notes

1. The present chapter has been collectively written by the team of the European Coordination Via Campesina (ECVC) in Brussels, and particularly Gérard Choplin (a member of the ECVC in Brussels), in collaboration with Jacques Berthelot (agricultural economist, member of the association Solidarity, France), Christian Boisgontier (former member of the Office of the CPE and a member of Economic and Social Committee, France), Guy Kastler (a member of the Peasant Confederation and the group of seeds of the ECVC, France), René Louail (former member of the ECVC committee and Regional Parliamentarian of Bretagne, France), Paul Nicholson (alternate member of the International Committee of Via Campesina, Pais Vasco, Spain), Josie Riffaud (member of the ECVC committee and the International Committee of Via Campesina, France), Geneviève Savigny (member of the ECVC committee, France), and Joan Verlinden (member of the ECVC committee, Belgium), among other contributors.
2. As well as the farms, regardless of their size, which are indebted and driven into a productivism spiral.
3. By Patrick Viveret (radio interview on France Inter, 18 September 2010, 'Parenthèse').
4. For example: cancellation of debts, contribution of capital to the consolidation of operations, quantity premiums, etc.
5. See the joint statement of ROPPA and Via Campesina (May 2001), available at www.ourworldisnotforsale.org/en/node/795.
6. Plants can directly extract nitrogen from the air, whereas manufacturing chemical nitrogen fertiliser consumes about a quarter of the energy used in agriculture in Europe.
7. At the international level, most members join the International Federation of Agricultural Producers (IFAP).
8. For more details, see the Via Campesina website, www.viacampesina.org.
9. See www.terredeliens.org.
10. Available at www.europeanfooddeclaration.org.

CONCLUSION

Facing the Domination of Financial Capital:
The Convergence of Peasant Struggles Today

Rémy Herrera and Kin Chi Lau

All the contributions to this book, whether they are theoretical or empirical, and whichever country or region they consider, emphasise the general failure of capitalism to solve agrarian and agricultural problems. The recent deterioration in the circumstances of peasant agriculture following the exacerbation of the food dimension of the current systemic crisis of capitalism has revealed and confirmed once again the permanent and structural inability of such a system to resolve the deep internal contradictions it has generated since its very origins, not only at the local, national and regional levels, but also worldwide.

Even in the richest countries of the North, where productivity boosted by technological progress is very high and food provision is available for a large majority of the population, the problems experienced by most family farms to keep their smallholdings and maintain their productive activities under satisfactory and decent working conditions, as well as the problems faced by consumers to master both the variety and the quality of their food, and indeed by every citizen to conserve natural resources and protect the environment, are exceeding the bounds of the bearable.

In the South – Latin America, Africa, Asia, Oceania – where average levels of productivity and mechanisation in agriculture are often weaker, the difficulties are even more worrying. Today, nearly half the Southern countries have lost the capacity to produce and supply what their people need to eat. Post-independence Africa was self-sufficient for its food provisioning at the beginning of the 1960s but is today the continent is a net food importer. As we write this, around three billion under-nourished persons – mostly poor peasants or landless people – are suffering from hunger, while vast numbers of rural families who have lost their lands no longer have access to the means of food production. In most peripheral societies pauperisation is spreading, and the living conditions in rural areas – as well as in the huge urban slums congested by the rural exodus – are simply inhuman and unacceptable.

Clearly identified by all the authors, the common enemy of the people – wherever they may be living (or just surviving), working and resisting in the

South or in the North – is financial capital, which pushes people deeper and deeper into indebtedness and consequently subjects them to super-exploitation. Despite the numerous, multidimensional and complex contradictions of the current world system, it is precisely high financial capital, in crisis, that has launched a modern *conquista*, characterised by repeated assaults on all public goods and the common heritage of humanity, through a commodification of life, including land and the environment, and an attack on livelihood, along with an over-exploitation of labour – peasants and workers taken as a whole.

As finance capitalism becomes more barbaric and destructive than ever, the structural problem for the survival of late capitalism is the downward pressure on rates of profit. Financialisation as an answer creates a debt-driven economy, and the only thing that this system will offer, until it is in its death agony, is the worsening exploitation of labour and life. The peasantries of the global South will continue to be dispossessed of their land and means of livelihood. The contradictions of the capitalist global system have now become so deep and so unsolvable that the system is on the verge of collapse. To be able to relaunch a cycle of expansion at the centre of the world system, the current systemic crisis must destroy gigantic amounts of fictitious capital and transfer the costs to the global South – to the majority of the world's population – as well as to the environment.

The present situation resembles, not the beginning of the end of the crisis, but rather the beginning of a long-running process of implosion and collapse of the present phase of financialised capitalism. For humanity to escape from this impasse, radical change is the only hope. This forces us to reconsider the alternatives of social transformation which must lie beyond capitalism.

The difficulties are significantly complicated by the choices made by most of the states in the global South – not only in the so-called 'emerging' countries, such as China, India, Brazil and South Africa, but also in the current 'revolutionary' processes of Latin America – in favour of (one of the many varieties of) capitalism. Looking beyond their recent successes in terms of high GDP growth rates, and despite their differentiated contents and implications, such pro-capitalist development strategies – including those implemented in China – are illusory and unsustainable.

Hence, for the great majority of the people in the South and in the North, the struggle against deteriorating conditions is at the same time the struggle against processes of the globalisation of capitalistic relations spearheaded by financial capital, that is, a struggle against capitalism itself, waged on multiple fronts. Among the programmatic demands are what La Via Campesina has campaigned for: agriculture should be withdrawn from the WTO; agro-fuels should be banned; and control of technology, pricing and market by transnational agribusiness corporations should be rejected. Demands put to the state to defend national food sovereignty are legitimate and necessary. However, it has to be reckoned that in the era of the hegemony of global capital and transnational business, the role of the state, more often than not, is compromised. Financial capital has forged interest blocs across local, national and international levels. Thus, exerting public pressure for critical policies against the aggression and manipulation of financial capital and transnational agribusiness is a necessary strategic move for mobilisation. While it needs to be stressed that a state's *raison d'être* is to protect society, failing which it might as well not exist, people and movements need at the same time to do everything in their power to reduce their dependency on capital, debt and the market. This is all the more necessary for peasant and family agriculture.

The guiding principle is community's control over and management of land and water as commons, which must not be allowed to be privatised or commodified. The struggles over water and gas in Cochabamba (2000) and La Paz (2005) are exemplary (Herrera 2010b). Agrarian reform to redistribute 'land to the tiller' is high on the agenda in most countries in South and South-East Asia, Africa and Latin America. As La Via Campesina demands, the struggle is not just for 'land' (for individual households to operate in an atomised manner, vulnerable to the dictates of the market and financial capital), but also for 'territory', which involves cultural, social and economic reorganisation of communal relations to produce and live in a cooperative or collective manner. This necessitates that the 'commons' are not objects for appropriation or control still operating within the logic of capitalism, but focal nodes supporting a different relationship of community members amongst themselves and with nature.

Food sovereignty remains at the core of the struggle. To attain food sovereignty, another mode of production has to be practised – one that is different from the capitalist mode of production dictated by speculative markets and extensive and intensive machines in the subjugation and expropriation of the people. This even calls into question national boundaries, for sustainable food production, distribution and consumption is based on bioregions and watershed systems rather than the political borders of modern nation states. What is also called into question is the mode of consumption and circulation, which can have such a destructive impact on nature and on the value systems of communities that through the centuries have acquired the wisdom to live in sustainable ways. One important insight concerns the value of sharing – a practice that goes beyond the monetary measures that reduce social relations to calculations of gain and loss. The people's struggles and demands show that in relating to each other, what needs to be envisioned includes modes other than those of capitalistic relations. They also demonstrate the importance of the ecological dimension by recognising that the current capitalist crisis is at the same time a profound ecological crisis brought about by the extractive industries that exhaust the earth's resources and contaminate water, land and air; the indus-trialisation that contributes to global warming and climate change; science and technology, on which modern capitalism thrives, and which have demonstrated their powers of overwhelming destruction not only through nuclear weapons (which are produced for intended mass annihilation) but also nuclear power plants such as the ones at Chernobyl and Fukushima (which bring unintended self-destruction); and the capitalistic systems of food production and supply that are completely dependent on petrol as fuel.

Thus, strategies must be devised for reducing dependence on or control by finance capitalism, ranging from establishing the state's control over financial capital to the protection of food and livelihood items from price speculation and market manipulation. For the social movements, the paramount task is to defend food sovereignty, not only at the national level but also at the local level. Local self-organisation at the grassroots level to assign high priority to food sovereignty and environmental security and to fend off attempts at manipulation by financial capital (even microcredit at the grassroots level is dubious in that it uses debt to control the peasants' mode of life and mode of production) requires direct action that is innovative in its intellectual and affective dimensions to go beyond the dead end of capitalism. In this connection, we see more and more debates in the social movements on the defence of the commons, re-ruralisation, re-peasantisation and rebuilding of rural and urban communities that nurture and practise values different from capitalistic ones – values of reciprocity and communality.

A radical re-imagination of the ways in which human societies produce and consume is the only way out of the current catastrophic crisis that humanity is in. Without food sovereignty, that is, autonomous communal self-management in the production, distribution and consumption of food, no sustainable, diversified economy or political autonomy will be built. Without reversing the logic of the maximisation of profit and the concentration of private ownership, especially that of land and the means of production, no state policy or leadership will be consistent or effective. Without radically questioning the hyper-concentration of power in the hands of high financial capital, no genuinely substantive democracy – one that embodies social progress as well as the participation of the people at all levels and in all the processes of decision making concerning their collective future – will be possible.

Thus, a key question before us is the question of subjectivity and agency, that is, the question of the production of subjectivities by the struggling people themselves in going beyond the contradictions that inform their struggles. How can we envisage the classes and the masses for this social transformation or revolution? What can be the role of family farmers, small peasants and farm workers? Many progressive movements and leftist thinkers have historically encountered ideological difficulties in understanding the peasantries and political difficulties in building class alliances with them. This has been and still is the case in most capitalist countries, even during revolutionary processes, even when peasants have been fundamental components of or actors in these revolutions, such as in France (1789), Mexico (1910), Russia (1917), China (1949) or Cuba (1959), besides others.

Yesterday as today, peasant and family agriculture is sometimes stereotyped as being underproductive, inefficient, backward, even archaic, and inevitably condemned, therefore, to disappear in the very movement towards 'development'. 'Modernisation' is too often conceived as (and reduced to) industrialisation, and more recently as extending services, that is, as being antagonistic to maintaining small- or medium-sized family agriculture that is oriented towards self-sufficiency and local demand. This amounts to saying that, notwithstanding the structural connections between modernisation, colonisation and racism, modernisation is a good thing to pursue and a *telos* to achieve.

Consequently, and unfortunately, the anti-capitalist nature of family agriculture is unheeded, and so its potential ability to trigger structural changes and transformation of the societies and economies we are living in is underestimated. In social movements or workers' organisations, many leftist theoreticians still feel that peasants are 'residuals' of the past, defending corporatist or sectoral interests, and they are not seen as fighting for common objectives that are convergent with those of other workers and citizens. For this to change, it is necessary to take a radical critique of a modernisation in which urbanisation and industrialisation have been presented as progress and development, the violence and plunder of imperialism and colonialism have been concealed or understated, and racism has been brought in to justify the pillage. Alongside this progress and development, which privileges science and technology and an anthropocentric exploitation of nature, what used to be the commons are seized from the users, especially food producers in rural and indigenous communities.

In this predatory onslaught on the commons, production, rather than for the reproduction and enhancing of lives, is put into service for the accumulation of more and more money – capital that seeks to command labour power and take control over every aspect of social life through the mechanisms and processes of privatisation. Thus, the processes of globalisation of capitalistic relations can be looked upon as the spread of cancerous cells traversing the entirety of social life. Exploitation takes place indiscriminately by subsuming every form of labour into the valorisation machine that produces values through the domination of fantasies and desires in the presence of an overflowing supply of monetary garb – the symbol of wealth and well-being that is in fact the instrument of the exploitation of life.

Hence, the struggle to recover the commons is an assertion of the right to autonomous life and self-management for the majority across the wide global spectrum. In the face of the difficult task of offsetting the almost irreversible damages to the very existence of the earth as habitat for humans and other species

– under global warming, climate change, and human-induced catastrophes like the nuclear crisis – farmers, as much as workers from other social sectors, are the protagonists and actors for change. Only an alliance of struggles on all fronts, building interdependent and mutual support as well as learning from one another, can enhance our capacities for autonomous life and self-management.

Access to land and other resources necessary for the reproduction of life, as commons, is a legitimate right for all peasants, workers and common people. If food sovereignty is to safeguard modes of autonomous collective self-management, it is necessary to accept the continuation of family agriculture in the foreseeable future in the twenty-first century. If agrarian and agricultural questions are to be solved, it will be obligatory to liberate ourselves from the destructive logic that currently drives capitalism under high-finance domination. If the present rules of the imperialist domination of international trade are to be modified, we – peasants, workers and people of the North and the South – must unite and together face our common enemies – financial capital and its local allies – in order to recreate viable visions, rebuild alternative strategies and participate in the long arduous road to communism.

REFERENCES

Action Aid (2007), 'The World Bank and Agriculture: A Critical Review of the World Bank's World Development Report'. Action Aid discussion paper, London.

Alden Wily, L. (2008), 'Whose Land is It? Commons and Conflict States. Why the Ownership of the Commons Matters in Making and Keeping Peace'. Rights and Resources Initiative, Washington, D.C.

Alternative Survey Group (various years), *Alternative Economic Survey*, Libéralisation sans Social Justice. Rainbow Publishers, Delhi.

Amanor, K. (2008), 'Sustainable Development, Corporate Accumulation and Community Expropriation: Land and Natural Resources in West Africa'. In Kojo S. Amanor and S. Moyo (eds), *Land and Sustainable Development in Africa*. Zed Books, London and New York.

Amarshi, A., K. Good and R. Mortimer (1979), *Development and Dependency: The Political Economy of Papua New Guinea*. Oxford University Press, Melbourne.

Amin, S. (1973), *Neocolonialism in West Africa*. Penguin Books, Harmondsworth.

—— (1974), *Unequal Development*. Monthly Review Press, New York.

—— (1977), *Imperialism and Unequal Development*. Monthly Review Press, New York.

—— (1978), *The Law of Value and Historical Materialism*. Monthly Review Press, New York.

—— (1980), *Class and Nation: Historically and in the Current Crisis*. Monthly Review Press, New York.

—— (1997), *Capitalism in the Age of Globalization*. Zed Books, London.

—— (1998), *Spectres of Capitalism, a Critique of Current Intellectual Fashions*. Monthly Review Press, New York.

—— (ed.) (2005), *Les luttes paysannes et ouvrières face aux défis du XXIᵉ siècle*. Les Indes savantes, Paris.

—— (2008), *The World We Wish to See: Revolutionary Objectives in the Twenty-First Century*. Monthly Review Press, New York.

—— (2011), *Le Monde arabe dans la longue durée: Le « printemps » arabe ?* Le Temps des Cerises, Paris.

—— (2013a), *The Implosion of Contemporary Capitalism*. Monthly Review Press, New York.

—— (2013b), 'China 2013', *Monthly Review*, vol.64, no.10, pp.14–33, March.

Anderson, T. (2006), 'On the Economic Value of Customary Land in Papua New Guinea', *Pacific Economic Bulletin*, vol.21, no.1, pp.138–52.

Antheaume, B., J. Bonnemaison, M. Bruneau and C. Taillard (1995), *Asie du Sud Est: Océanie*. Belin-Reclus, Paris.

Arrighi, G. (1973), 'International Corporations, Labour Aristocracies, and Economic Development in Tropical Africa', in G. Arrighi. and J. Saul (eds), *Essays on The Political Economy of Africa*. Monthly Review Press, New York.

Bank of Papua New Guinea (various years), *Quarterly Economic Bulletin*. Port Moresby.

Banks, G. and C. Ballard (eds) (1997), *The Ok Tedi Settlement: Issues, Outcomes, and Implications*. National Centre for Development Studies, Australian National University, Canberra.

Berg, R. (1981), 'Accelerated Development in Sub-Saharan Africa: An Agenda for Action'. World Bank, Washington, D.C.

Bernstein, H. (2002), 'Agrarian Reform after Developmentalism?', presentation at the conference on 'Agrarian Reform and Rural Development: Taking Stock', Social Research Centre of the American University in Cairo, 14–15 October 2001.

Berthelot, J. (2001), *L'Agriculture, talon d'Achille de la mondialisation : Clés pour un accord agricole solidaire à l'OMC*. L'Harmattan, Paris.

—— (2008), 'Analyse critique des causes essentielles de la flambée des prix agricoles mondiaux'. Association Solidarité, Toulouse.

—— (2012), 'Libres échanges avec Jacques Berthelot : les Ape vont appauvrir encore plus les pays d'Afrique', available at www.lequotidien.sn/index.php/economie/item/16579.

Bird, K., D. Booth and N. Pratt (2002), 'The Contribution of Politics, Policy Failures and Bad Governance to Food Security Crisis in Southern Africa', theme paper commissioned by the Forum for Food Security in Southern Africa, available at www.odi.org.uk/food-security-forum.

Bourke, R.M. (2001), 'Intensification of Agricultural Systems in Papua New Guinea', *Asia Pacific Viewpoint*, vol.42, no.2/3, pp.219–35.

—— (2005), 'Agricultural Production and Customary Land in Papua New Guinea', in J. Fingleton (ed.), *Privatising Land in the Pacific*, no.80. Australia Institute, Canberra.

—— (2009), 'History of Agriculture in Papua New Guinea', in R.M. Bourke and T. Harwood (eds), *Food and Agriculture in Papua New Guinea*, pp.10–23. Australian National University, ANU E-press, Canberra.

Bourke, R.M. and V. Vlassak (2004), *Estimates of Food Crop Production in Papua New Guinea*. Australian National University, Canberra.

Braudel, F. (1981), *Grammaire des civilisations*. Flammarion, Paris.

—— (1986), *L'Identité de la France: Espace et histoire*. Arthaud-Flammarion, Paris.

Braun, J. von and R. Meinzen-Dick (2009), 'Land Grabbing by Foreign Investors in Developing Countries: Risks and Opportunities', IFPRI Policy Brief No.13, April. IFPRI, Washington, D.C.

Bruce, J.W. (1988), 'A Perspective on Indigenous Land Tenure Systems and Land Concentration', in R.E. Downs and S.P. Reyna (eds), *Land and Society in Contemporary Africa*, pp.23–52. University Press of New England, Hanover and London.

Caldart, R.S. (2000), *Pedagogia do Movimento Sem Terra: Eescola e Mais do que Escola*, 2nd edn. Editora Vozes, Petrópolis.

—— (2006), 'Movement of the Landless Rural Workers (MST): Pedagogical Lessons', in E. Vieira (ed), *The Sights and Voices of Dispossession: The Fight for the Land and the Emerging Culture of the MST in Brazil*. School of Modern Languages, University of Nottingham, available at http://land-less-voices. org/vieira.

Carcanholo, R.A. (2014), 'The Great Depression of the 21st Century and Fictitious Wealth: On the Theoretical Categories of Fictitious Capital and Fictitious Profit', in Herrera, Dierckxsens and Nakatani, pp.153–75.

Casanova, A. and R. Herrera (eds) (2014), *Penser les crises*. Le Temps des Cerises, Paris.

Ceceña, A.E. (2004), 'Geographies of the *Zapatista* Uprising', *Antipode*, vol.36, no.3, pp.391–9.

CETIM and GRAIN (2012), *Hold-up sur l'alimentation : comment les sociétés transnationales contrôlent l'alimentation du monde, font main basses ur les terres et détraquent le climat*. Centre Europe – Tiers Monde, Geneva.

Chandrasekhar, C.P. (2011), 'No Jobs out There', *The Hindu*, 2 July.

Chayanov, A.V. (1986), *The Theory of Peasant Economy*. Manchester University Press, Manchester.

Collier, P. (2007), *The Bottom Billion: Why the Poorest Countries Are Failing and What Can Be Done About It*. Oxford University Press, New York.

Cooper, F. (1994), 'Conflict and Connection: Rethinking Colonial African History', *American Historical Review*, vol.99, no.5, pp.1516–45.

Corrin Care, J.C. and D.E. Paterson (2008), *Introduction to South Pacific Law*. Routledge Cavendish, Abingdon.

Cotula, L., S. Vermeulen, R. Leonard and J. Keeley (2009), *Land Grab or Development? Agricultural Investments and International Land Deals in Africa*, IIED, FAO and IFAD, available at www.fao.org/3/a-ak241e.pdf.

Crédit Suisse (2013), *Global Wealth Report 2013*. Research Institute, Zurich, October.

Crocombe, R. (1987), *Land Tenure in the Pacific*. University of the South Pacific, Suva.

Curtin, T. and L. David (2006), 'Land Titling and Socioeconomic Development in the South Pacific', *Pacific Economic Bulletin*, vol.21, no.1, pp.153–80.

Dale, P.F. (1976), *Cadastral Surveys within the Commonwealth*, Overseas Research Publication no.23. Her Majesty's Stationery Office, London.

Defert, G. (1996), *L'Indonésie et la Nouvelle-Guinée occidentale: maintien des frontières coloniales ou respect des identités communautaires*. L'Harmattan, Paris.

Delcourt, L. (2010), 'L'Avenir des agricultures paysannes face aux nouvelles pressions sur la terre', *Alternatives Sud*, vol.17, no.3, pp.7–34.

Delhi Science Forum (various years), *The Indian Economy: An Alternative Survey*. Prajasakti Book House, Hyderabad.

Denham, T.P., S.G. Haberle and C. Lentfer (2004), 'New Evidence and Revised Interpretations of Early Agriculture in Highland New Guinea', *Antiquity*, no.78, pp.839–57.

—— —— ——, R. Fullagar, J. Field, M. Therin, N. Porch and B. Winsborough (2003), 'Origins of Agriculture at Kuk Swamp in the Highlands of New Guinea', *Science*, no.301, pp.189–93.

Department of Mining (various years), *Information Booklets*, Port Moresby.

Dorward, A.R, J.F. Kirsten, S.W. Omamo, C. Poulton and N. Vink (2009), 'Institutions and the Agricultural Development in Africa', in J.F. Kirsten, A.R. Dorward, C. Poulton, and N. Vink (eds), *Institutional Economics Perspectives on African Agricultural Development*, pp.3–34. IFPRI, Washington, D.C.

Duby, G. and A. Wallon (1976), *Histoire de la France rurale*, 4 vols. Édition du Seuil, Paris.

Economist Intelligence Unit (various years), 'Country Profile'. *The Economist*, London, available at www.eiu.com.

Evenson, R.E. and P. Pingali (eds) (2007), *Handbook of Agricultural Economics – Agricultural Development: Farmers, Farm Production and Farm Markets*, vol.3. Elsevier, North-Holland, Amsterdam.

FAO (1983), *Approaches to World Food Security*. Food and Agricultural Organization of the United Nations, Rome.

—— (2006), 'Investir dans l'agriculture pour endiguer l'exode rural', *Étude sur le role de l'agriculture*. Food and Agriculture Organization of the United Nations, Rome.

—— (2013), *The State of Food and Agriculture 2013*. Food and Agriculture Organization of the United Nations, Rome.

—— et al. (2011), 'Price Volatility in Food and Agricultural Market: Policy Responses', *Background Policy Report for the G20 Summit in Paris in November*, Rome.

Fews Net (various years), *Food Security Outlooks*, Famine Early Warning Systems Network, available at www.fews.net/.

Filer, C. (2006), 'Custom, Law and Ideology in Papua New Guinea', *Asia Pacific Journal of Anthropology*, vol.7, no.1, pp.65–84.

—— , J. Burton and G. Banks (2008), 'The Fragmentation of Responsibilities in the Melanesian Mining Sector', in C. O'Faircheallaigh and S. Ali (eds), *Earth Matters: Indigenous Peoples, the Extractive Industries, and Corporate Social Responsibility*, pp.163–179. Greenleaf Publishing, London.

—— and B. Imbun (2009), 'A Short History of Mineral Development Policies in Papua New Guinea: 1972–2002', in R.J. May (ed.), *Policy Making and Implementation: Studies from Papua New Guinea*, pp.75–116. Australian National University, ANU E-press, Canberra.

Fingleton, J. (1991), 'The East Sepik Land Legislation', in P. Larmour (ed.), *Customary Land Tenure: Registration and Decentralisation in Papua New Guinea*, Monograph no.29. National Research Institute, Port Moresby.

—— (2007), 'A Legal Regime for Issuing Group Titles to Customary Land: Lessons from the East Sepik', in J. Weiner and K. Glaskin (eds), *Customary Land Tenure and Registration in Australia and Papua New Guinea: Anthropological Perspectives*. Australian National University, ANU E-press, Canberra.

Foley, W.A. (1986), *The Papuan Languages of New Guinea*. Cambridge University Press, Cambridge.

Frank, Andre Gunder (1969), *Latin America: Underdevelopment or Revolution*. Monthly Review Press, New York.

GeneWatch UK and Greenpeace International (2008), 'GM Contamination Register Annual Report', available at http://tinyurl.com/79osjp.

Gewertz, D.B. and F.K. Errington (1999), *Emerging Class in Papua New Guinea: The Telling of Difference*. Cambridge University Press, Cambridge.

Ghosh, J. (2008), 'The Global Oil Price Story', available at www.networkideas.org/jul2008/ news28_oil_price.htm.

Gille, B. and P.Y. Toullelan (1999), *De la conquête à l'exode*. Au Vent des Îles, Paris.

Godelier, M. (1982), *La Production des grands hommes: Pouvoir et domination masculine chez les Baruya de Nouvelle Guinée*. Fayard, Paris.

Golson, J. (1991), 'The New Guinea Highlands on the Eve of Agriculture', *Bulletin of the Indo-Pacific Prehistory Association*, no.11, pp.82–91.

Glantz, M.H., M. Betsil and K. Crandall (2007), 'The 1991/92 Drought: Historical Context', in *Food Security in Southern Africa: Assessing the Use and Value of ENSO Information*, University Corporation for Atmospheric Research (UCAR), available at www.isse.ucar.edu/sadc/chptr4.html.

GRAIN (2007), *Seedling*, Agrofuels Special Issue, GRAIN Report, available at www.mstbrazil.org/grainsaysnoagrofuels.

—— (2009), 'Grabbing Land for Food', *Grain Seedling*, January.

—— (2014), 'Food Sovereignty for Sale: Supermarkets are Undermining People's Control over Food and Farming in Asia', Barcelona, available at www.grain.org.

Hardt, M. and A. Negri (2009), *Commonweatlh*. The Belknap Press of Harvard University Press, Massachusetts.

Harvey, D. (2010), *A Companion to Marx's Capital*. Verso, London and New York.

Herman, E. and N. Chomsky (1988), *Manufacturing Consent: The Political Economy of the Mass Media*. Pantheon, New York.

Herrera, R. (2006), 'The "New" Development Economics: A Neoliberal Con?', *Monthly Review*, vol.58, no.1, pp.38–50.

—— (2010a), *Un Autre Capitalisme n'est pas possible*. Syllepse, Paris.

—— (2010b), *Les Avancées révolutionnaires en Amérique latine: Des Transitions socialistes au XXIe siècle?* Parangon, Lyon.

—— (2011), 'A Critique of Mainstream Growth Theory: Ways out of the Neoclassical Science(-Fiction) and Towards Marxism', *Research in Political Economy*, vol.27, no.1, pp.3–64, New York.

—— (2012), 'Reflections on the Current Crisis and its Effects', *Economic and Political Weekly*, vol.47, no.23, pp.62–71, Mumbai.

—— (2013a), 'Neoclassical Economic Fiction and Neoliberal Political Reality', *International Critical Thought*, vol.3, no.1, pp.98–107, London.

—— (2013b), 'Between Crisis and Wars: Where Is the United States Heading?', *Journal of Innovation Economics and Management*, no.12, 2013/2, pp.151–74, Brussels.

—— Dierckxsens, W. and P. Nakatani (eds) (2014), *Beyond the Systemic Crisis and Capital-led Chaos: Theoretical and Applied Studies*, P.I.E. Peter Lang, Brussels and Berlin.

—— and P. Tetoe (2011), 'Le Capital étranger dans le secteur des mines en Papouasie Nouvelle-Guinée', *Économies et Sociétés*, série F, vol.45, no.2, pp.321–6.

—— —— (2012a), 'Le Paradoxe *Papua Niugini*: archaïsme de la propriété de la terre, modernité des résistances paysannes?', *Revue française de Socio-économie*, no.9, 2012/1, pp.133–52.

—— —— (2012b), 'La Papouasie Nouvelle-Guinée entre dépendance et résistances', *La Pensée*, no.369, pp.97–115.

—— —— (2012c), 'La Papouasie Nouvelle-Guinée chez elle ou l'ambivalence des relations avec l'Australie', *Journal de la Société des Océanistes*, vol.2, no.135, pp.201–14.

Holt-Giménez, E. (2006), *Campesino a Campesino: Voices from Latin America's Farmer to Farmer Movement for Sustainable Agriculture*. Food First Book, Oakland, Calif.

Hope, G.S. and S.G. Haberle (2005), 'The History of the Human Landscapes of New Guinea', in A. Pawley, R. Attenborough, J. Golson and R. Hide (eds), *Papuan Pasts*, pp.541–54. Australian National University, Pacific Linguistics, Canberra.

Houtart, F. (2009), *Agrofuels*. Pluto Press, London.

Iatau, M. and I. Williamson (1997), 'An Introduction to the Use of Case Study Methodology to Review Cadastral Reform in Papua New Guinea', Commission 7 Symposium, 64th Permanent Committee of the International Federation of Surveyors, Singapore.

Imbun, B. (2000), 'Mining Workers or "Opportunist Tribesmen"? A Tribal Workforce in a Papua New Guinea Mine', *Oceania*, no.71, pp.129–49.

James, R.W. (1985), *Land Law and Policy in Papua New Guinea*, Monograph no.5. Law Reform Commission of Papua New Guinea, Port Moresby.

Kalibwani, X. (2005), 'Food Security in Southern Africa: Current Status, Key Policy Processes and Key Players at Regional Level', background paper for 'Promoting the Use of CSO Evidence in Policies for Food Security: An Action Research Project in Southern Africa', ODI/SARPN/FANPRAN, October.

Kanyinga, K. (2000), 'Re-distribution from Above: The Politics of Land Rights and Squatting in Coastal Kenya', a report from the research programme 'The Political and Social Context of Structural Adjustment in Africa', no.115. Nordiska Afrikainstitutet, Uppsala.

Kanyongolo, F.E. (2005), 'Land Occupations in Malawi: Challenging the Neoliberal Legal Order', in S. Moyo and P. Yeros (eds), *Reclaiming the Land: The Resurgence of Rural Movements in Africa, Asia and Latin America*. Zed Books, London.

Kautsky, K. (1988), *On the Agrarian Question* (1899), trans. P. Burgess. Zwan Publications, London and Winchester, Mass.

Kavanamur, D. (1997), 'The Politics of Structural Adjustment in Papua New Guinea', in P. Larmour (ed.), *Governance and Reform in the South Pacific*. Australian National University, Canberra.

Kong, Xiangzhi and He, A. (2009), 'The Contribution of Peasants to Nation Building in the First 60 Years of the People's Republic of China', *Teaching and Research*, no.9. ('新中国成立60年来农民对国家建设的贡献分析', 教学与研究).

Lalau, A.A. (1991), 'State Acquisition of Customary Land for Public Purposes in Papua New Guinea', Technical Report Series no.91–1. Department of Surveying and Land Studies, Papua New Guinea University of Technology, Lae, Morobe Province.

Larmour, P. (2003), 'Land Registration in Papua New Guinea: Competing Perspectives', discussion paper. Research School of Pacific and Asian Studies, Australian National University, Canberra.

Latour, B. (1993), *We Have Never Been Modern*. Harvard University Press, Cambridge, Mass.

Lefort, C. (1986), *The Political Forms of Modern Society*. Polity Press, UK.

Lenin, V.I. (1965), *Collected Works*. Progress Publishers, Moscow, available at Marxist Internet Archives, www.marxists.org/archive/lenin/works/.

Liang, S. (1977), 'Memories and Reflections of My Engagement with the Rural Reconstruction Movement', *Collections of Liang Shuming's Work*, vol.7, pp.424–8.

—— (2006), *Theory of Rural Reconstruction: The Future of the Chinese Nation* [1937] (乡村建设理论, 一名中国民族之前途). Shanghai: Century Publisher.

Mafeje, A. (1999), 'Imperatives and Options for Agricultural Development in Africa: Peasant or Capitalist Revolution?' (unpublished).

—— (2003), *The Agrarian Question, Access to Land and Peasant Responses in Sub-Sahara Africa*, UNRISD programme papers on Civil Society and Social Movements, Geneva.

Mamdani, M. (1996), *Citizens and Subjects: Contemporary Africa and the Legacy of Late Colonialism*. Princetown University Press, UK.

Manji, A. (2006), *The Politics of Land Reform in Africa: From Communal Tenure to Free Markets*. Zed Books, London and New York.

Marx, K. (1881), 'First Draft of Letter to Vera Zasulich', available at www.marxists.org/archive/marx/works/1881/03/zasulich1.htm.

—— (1973), *Grundrisse*. Penguin Books, Harmondsworth, also available at www.marxists.org /archive/marx/works/1857/grundrisse/.

—— (1974), *Capital*, Volume 3, ed. F. Engels. Progress Publishers, Moscow.

—— (1976), *Capital: A Critique of Political Economy*, translated by Ben Fowkes. Penguin Books, Harmondsworth.

—— and F. Engels (1882), *The Communist Manifesto*, Russian edition, available at www.marxists.org/archive/marx/works/1848/communist-manifesto/preface. htm#preface-1882.

Mazoyer, M. (2002), 'Une Situation agricole mondiale insoutenable, ses causes et les moyens d'y remédier', *Mondes en développement*, vol.30, no.117, pp.725–37.

—— and L. Roudart (1997), *Histoire des agricultures du monde: du néolithique à la crise contemporaine*. Le Seuil, Paris.

Migot-Adholla, S.E. (1994), 'Land, Security of Tenure and Productivity in Ghana', in J.W. Bruce and S.E. Migot-Adholla (eds), *Searching for Land Tenure Security in Africa*, pp.169–98. Kendall Hunt Publishing, Dubuque.

Miller, J.W. (2010), 'World Bank Land Grab Report Under Fire', *Wall Street Journal*, 29 July.

Ministry of Statistics and Programme Implementation (various years), *Key Indicators of Employment and Unemployment in India*. National Sample Survey Office (NSSO), New Delhi.

Mkandawire, R. and K. Matlosa (eds) (1993), *Food Policy and Agriculture in Southern Africa*. SAPES Books, Harare.

Mkandawire, T. and C. Soludo (1999), *Our Continent, Our Future: African Perspectives on Structural Adjustment*. CODESRIA, IDRC and AWP, Canada.

Moyo, S. (2008), *African Land Questions, Agrarian Transitions and the State: Contradictions of Neoliberal Land Reforms*, CODESRIA Green Book Series. CODESRIA, Dakar.

—— and P. Yeros (2005), 'The Resurgence of Rural Movements under Neoliberalism', in S. Moyo and P. Yeros (eds), *Reclaiming the Land: The Resurgence of Rural Movements in Africa, Asia and Latin America*. Zed Books, London.

—— —— (forthcoming), 'After Zimbabwe: State, Nation and Region in Africa', in S. Moyo, P. Yeros and J. Vadell (eds), *The National Question Today: The Crisis of Sovereignty in Africa, Asia and Latin America*. Pluto Press, London.

Nagaraj, K. (2008), *Farmers' Suicides in India: Magnitudes, Trends and Spatial Patterns*, Preliminary Report. Madras Institute of Development Studies, Chenai, available at www.macroscan.org.

Nakatani, P. and R. Herrera (2010), 'Keynes (et Marx), la monnaie et la crise', *La Pensée*, no.364, pp.57–68.

—— (2013), 'Keynes et la crise: Hier et aujourd'hui', *Actuel Marx*, no.53, pp.153–68.

NBS (various years), *Statistical Database*. National Bureau of Statistics of the People's Republic of China, Beijing, available at www.stats.gov.cn/english/.

Neale, T. (2005), 'Historical Overview of Mining in PNG', *Melanesian Resources*, available at http://infomine.com/.

Ng, F. and A. Yeats (1996), 'Open Economies Work Better! Did Africa's Protectionist Policies Cause Its Marginalization in World Trade?', World Bank Working Paper No.1636, Washington, D.C.

Nicholson, P., X. Montagut and J. Rulli (2012), *Terre et liberté! À la conquête de la souveraineté alimentaire*. Centre Europe Tiers-Monde (CETIM), Geneva.

ÖBV (2010), Bäuerliche Zukunft, no.313, ÖBV (Via Campesina Austria), March, available at www.viacampesina.at/cm3/zeitung-abonnieren.html.

O'Connor, S. and J. Chappell (2003), 'Colonisation and Coastal Subsistence in Australia and Papua New Guinea: Different Timing, Different Modes?', in C. Sand (ed.), *Pacific Archaeology*, pp.17–32. Les Cahiers de l'Archéologie en Nouvelle-Calédonie, Nouméa.

Patnaik, P. (2003), *The Retreat to Unfreedom*. Tulika Books, New Delhi.

—— (2008), 'The Accumulation Process in the Period of Globalization', available at www.networkideas.org/focus/may 2008/fo28_globalisation.htm.

—— (2011), *Re-Envisioning Socialism*. Tulika Books, New Delhi.

Pawley, A., R. Attenborough, J. Golson and R. Hide (eds) (2005), *Papuan Pasts: Cultural, Linguistic and Biological Histories of Papuan-speaking Peoples*. Australian National University – Research School of Pacific and Asian Studies, Pacific Linguistics, Canberra.

Petras, J. (2008), The Great Land Giveaway: Neo-colonialism by Invitation, available at www.globalresearch.ca/the-great-land-giveaway-neo-colonialism-by-invitation/ 11231.

Polanyi, K. (1944), *The Great Transformation*. Rinehart, New York.

Power, T. (1991), 'Policy Making in East Sepik Province', in P. Larmour (ed.), *Customary Land Tenure: Registration and Decentralisation in Papua New Guinea*, monograph no.29. National Research Institute, Port Moresby.

Reserve Bank of India (various years), *Some Basic Statistics Relating to the Indian Economy*, Mumbai.

Roberts, S. (2008), 'Prices of Grain, Flour, Mealie Meal and Bread' (unpublished).

Rosegrant, M.W. (2008), 'Biofuels and Grain Prices: Impacts and Policy Responses', testimony for the US Senate Committee on Homeland Security and Governmental Affairs, Washington, D.C.

Schroeder, R. (1992), 'Initiation and Religion: A Case Study from the Wosera of Papua New Guinea'. University Press Fribourg Switzerland, Fribourg.

Scott, J.C. (1985), *Weapons of the Weak: Everyday Forms of Peasant Resistance*. Yale University Press, New Haven.

—— (1990), *Domination and the Arts of Resistance: Hidden Transcripts*. Yale University Press, New Haven.

SCTL (2002), 'La Société Civile des Terres du Larzac, une approche novatrice et originale de la gestion foncière des territoires ruraux', available at www.agter. asso.fr/article232_fr.html.

SEPA (various years), 'Site Statistics'. State Environmental Protection Administration, Beijing.

Shivji, I.G. (2009), *Where is Uhuru? The Struggle for Democracy in Africa*. Fahamu Books, Oxford.

Sibanda, A. (1988), 'The Political Situation', in Colin Stoneman (ed.), *Zimbabwe's Prospects: Issues of Race, Class, State and Capital in Southern Africa*. Macmillan Publishers, London and Basingstoke.

Social Network for Justice and Human Rights (2007), available at www. grassrootsonline.org/term/social-network-justice-and-human-rights.

Stedile, J.P. (2002), 'Landless Battalions: The *Sem Terra* Movement of Brazil', *New Left Review*, May–June, available at www.mstbrazil.org/stedileinterviewMST.

—— (2005), *A Questão Agrária no Brasil: O Debate Tradicional 1500–1960*, vol.1. Editora Expressão Popular, São Paulo.

—— (2006), *A Questão Agrária no Brasil: O Debate na Esquerda 1960–1980*, vol.2. Editora Expressão Popular, São Paulo.

—— (2007a), *A Questão Agrária no Brasil: Programas de Reforma Agrária 1946–2003*, vol.3, Editora Expressão Popular, São Paulo.

—— (2007b), 'The Neoliberal Agrarian Model in Brazil', *Monthly Review*, vol.58, no.8, pp.50–4.

—— and B.M. Fernandes (1999), *Brava Gente: A Trajetória do MST e a Luta pela Terra no Brasil*. Editora da Fundação Perseu Abramo, São Paulo.

—— and Frei S. Görgen (1993), *A Luta Pela Terra no Brasil*. Editora Página Aberta, São Paulo.

Stewart, P. and A. Strathern (eds) (2005), *Anthropology and Consultancy: Issues and Debates*, Studies in Applied Anthropology series. Berghahn Books, New York.

Strathern, M. (2009), 'Land: Intangible or Tangible Property?', in T. Chesters (ed.), *Land Right: The Oxford Amnesty Lectures 2005*, pp.13–38. Oxford University Press, Oxford.

Sullivan, M.E., P. Hughes and J. Golson (1987), 'Prehistoric Garden Terraces in the Eastern Highlands of Papua New Guinea', *Tools and Tillage*, no.5, pp.199–213.

Tabb, W.K. (2008), 'The Global Food Crisis and What Has Capitalism to Do With It', available at www.networkideas.org/focus/Jul2008/fo28_Global_Food_Crisis.htm (accessed on 28 July 2008).

Thompson, C. (2008), 'Bio-Fuels for Africa?', prepared for National Consultative Workshop on Current Issues affecting Agro-Biodiversity for Civil Society Positions to CBD-COP9, held at ZIPAM, 28–30 April, Norton, Zimbabwe.

Transnational Institute (2007), *Agrofuels: Towards a Reality Check in Nine Key Areas*. TNI Report, Amsterdam, available at www.mstbrazil.org/ transnationalinstituteon agrofuelsrealitycheck.

Trebilcock, M.J. (1983), 'Customary Land Law Reform in Papua New Guinea: Law, Economics and Property Rights in a Traditional Culture', *Adelaide Law Review*, vol.1, no.9, pp.191–228.

UNCTAD (2008), *UNCTAD Handbook of Statistics 2008*. United Nations, New York and Geneva.

UNDP (various years), *Human Development Report*, United Nations Development Programme, Geneva.

UNECA (2004), *Assessing Regional Integration in Africa*, ECA Policy Research Report. Economic Commission for Africa, Addis Ababa.

UNEP (2002), *Africa Environment Outlook: Past, Present and Future Perspectives*. Earthprint, UK.

UNICEF (2009), 'The State of the World's Children 2009', New York.

Via Campesina (2010), 'La Situation des paysans européens et leurs luttes', available at www.viacampesina.org/downloads/pdf/openbooks/FR-02.pdf.

Vitali, S., J.B. Glattfelder and S. Battison (2011), 'The Network of Global Corporate Control'. ETH, Zurich.

Wahenga (2007), 'Bio-fuel Production and the Threat to South Africa's Food Security', Wahenga Brief No.11, April, Regional Hunger Vulnerability Programme (RHVP), available at www.wahenga.net.

Watts, M. (2007), 'Chronology of Bougainville Civil War' (unpublished mimeo), Sydney.

Weiner, J.F. and K. Glaskin (eds) (2007), 'Customary Land Tenure and Registration in Indigenous Australia and Papua New Guinea: Anthropological Perspectives', Asia-Pacific Environment Monograph 3. Australian National University, ANU E-press, Canberra.

Wen, J. (no date) 'China's Agricultural and Rural Development', available at www.fao.org.

Wen, T. (2001), 'Centenary Reflections on the "Three Dimensional Problem" of Rural China', trans. P. Liu, *Inter-Asia Cultural Studies*, vol.2, no.2, pp.287–95 (originally printed in *Dushu*, no.12, pp.3–11, 1999).

—— (2009), *The 'San Nong' Problem and Institutional Transition* (三农问题与制度变迁). China Economic Press, Beijing.

—— X. Dong, S. Yang, X. Liu and K.C. Lau (2012), 'China Experience, Comparative Advantage, and the Rural Reconstruction Experiment', in A. Dirlik, R. Prazniak and A. Woodside (eds), *Global Capitalism and the Future of Agrarian Society*. Paradigm Publishers, London.

Young, R.J.C. (2004), *White Mythologies*, 2nd edn. Routledge, London.

Williamson, I.P. (1997), 'The Justification of Cadastral Systems in Developing Countries', *Geomatica*, vol.51, no.1, pp.21–36.

World Bank (various years), *World Development Report*. World Bank, Washington, D.C.

—— (1978), *PNG: Economic Situation and Prospects for Development*. World Bank, Washington, D.C.

—— (1989), 'Land Mobilisation Project', Report No.7592-PNG. Department of Lands, Port Moresby.

—— (2008), *World Development Report: Agriculture for Development*. World Bank, Washington, D.C.

—— (2013), *World Development Report: Jobs*, World Bank, Washington, D.C.

—— (2014), 'Gross National Income Per Capita Atlas Method and PPP', *World Development Indicators Database 2012*. World Bank, Washington, D.C., April.

Zhang, Yi (2005), 'The Mutilated Migrant Workers: A Bloodstained Story Behind the Economic Growth', *China Youth Daily*, 27 April.

CONTRIBUTORS

Samir Amin is the president of the WFA and the director of the Third World Forum. He is one of the founders of the capitalist world system theory.

Gérard Choplin is a member of the European Coordination Via Campesina in Brussels. His chapter was written in collaboration with Jacques Berthelot, Christian Boisgontier, Guy Kastler, René Louail, Paul Nicholson, Josie Riffaud, Geneviève Savigny and Joan Verlinden.

Rémy Herrera is a researcher at the CNRS (National Center for Scientific Research, France) and teaches at the University of Paris 1 Panthéon-Sorbonne. He is also executive secretary of the WFA.

Kin Chi Lau is associate professor at Lingnan University in Hong Kong, vice-president of the WFA and member of the Steering Committee of South South Forum on Sustainability (SSFS).

Sam Moyo director of the Centre for Agricultural Studies in Harare, the former president of the CODESRIA and a member of the WFA.

Utsa Patnaik is national fellow at the Indian Council of Social Science Research and (retired) professor of economics at the Jawaharlal Nehru University, New Delhi.

Jade Tsui Sit is associate professor at Southwest University in Chongqing, China, and a board member of ARENA (HK).

João Pedro Stedile is a member of the National Coordination Committee of the Movement of Landless Workers (La Via Campesina).

Poeura Tetoe is currently a Ph.D. student at the University of French Polynesia, Papetee, Tahiti.

Erebus Wong is a member of the Steering Committee of South South Forum on Sustainability and research coordinator of ARENA (HK).

INDEX

soil exhaustion, 68
Solidarity, 153
South Africa, 4, 7, 16, 24, 27, 30, 56,
   58, 60, 62, 64, 65, 66, 68, 69,
   71–7, 78, 80, 81, 141, 155;
   agribusiness, 73; capital, 77
South Korea, 63
Southern Africa, 62, 63, 65, 69, 77
Southern African Customs Union,
   73
Southern African Development
   Community, 7; agrarian crisis,
   77; agricultural export, 72; food
   crisis, 71; Free Trade Area, 74
Soviet Union, 16, 85, 107
soybean output, 141
Spain, 146
special economic zones, in India, 10,
   114, 117
speculation, on land, 73; on real
   estate, 111
speculative capital, 69
standardization, 39, 48
Star Mountains, 131
state control, 40
state expropriations of land, Africa,
   62
state ownership of land, in China,
   25
state policy, Africa, 60, 65, 67
Stedile, João Pedro, 5-6, 35–55
structural adjustment plans, 1, 7, 12,
   57, 62, 67, 71, 74, 75, 77, 128,
   131, 132; Africa, 57, 65
structural distortion, Africa, 65
struggles, against dam in Khandwa of
   Madhya Pradesh, 115; alliance,
   158; for land in Gujarat,
   Mumbai, Pune of Maharashtra,
   Uttar Pradesh, Delhi, 115; for
   territory, 48; for the commons,
   158-9

subalterns, 104
subjectivity production, 157
sub-Saharan Africa, 2, 76
Suez, 54
supermarketisation, 78
sustainability, 103; peasant
   agriculture, 151
Swaziland, 66, 72
Syngenta, 54, 147

Tan, Shuhao, 88–9, 107
Tanzania, 60, 66, 72, 77
technology, 137, 157, 158
Tee Cycle, 121
temporary employment, 43
Terre des Liens, 145
Tetoe, Poeura, 11–12, 119–35
Thailand, 70, 131
Tok Pisin, 134
trade agreements, EU, 150
trade liberalisation, 67, 69; Africa, 61
training, grassroots, 51
transnational corporations, 2, 11–12,
   35–6, 38–40, 46, 48–9, 54, 60,
   69, 119, 125, 131, 132, 137;
   Papua New Guinea, 133
trickle-down theory, 110
Tunisia, 24
Turkey, 141

Uganda, 59
unemployment, China, 88; India,
   110–13; EU, 137
United Kingdom, 136
United States, 14, 15, 16, 19, 20, 32,
   33, 34, 37, 43, 63, 74, 86, 92, 102,
   128, 129, 136, 140
USAID, 129, 134

Verlinden, Joan, 153
Vietnam, 70; land tenure, 25
Viveret, Patrick, 153